THE
YEAR
THE WORLD
WENT MAD

THE
YEAR
THE WORLD
WENT MAD

MARK WOOLHOUSE

Foreword by Matt Ridley

SANDSTONE PRESS

First published in Great Britain in 2022
Sandstone Press Ltd
PO Box 41
Muir of Ord
IV6 7YX
Scotland

www.sandstonepress.com

ISBN: 978-1-913207-95-3
ISBNe: 978-1-913207-96-0

Sandstone Press is committed to a sustainable future.
This book is made from Forest Stewardship Council® certified paper.

Cover design by Heike Schüssler
Typeset by Iolaire Typesetting, Newtonmore
Printed and bound by TJ Books Limited, Padstow, Cornwall

This book is for my wife and daughter, the best lockdown companions I could ever wish for.

Francisca, thank you for listening. You must have heard everything in this book a hundred times before.

Nyasha, thank you for bringing joy to our lockdown lives. I'm sorry that your generation has been so badly served by mine.

CONTENTS

FOREWORD

Mark Woolhouse is one of the world's most distinguished epidemiologists. His expertise has been invaluable to Scotland, Britain and the world during the Covid pandemic. His account of the first year of Covid is a remarkable story, told with great fluency and insight by somebody who was on the scientific and political inside throughout. But he is also frustrated and baffled by one big mistake that both the United Kingdom and the world made against his advice, so his book has a critical – in both senses of the word – argument to make.

That mistake was lockdown. Mark argues that however great the threat posed by the novel coronavirus, and however badly it was underestimated at first, there was always a better strategy to deal with the resulting epidemic that would have done far less economic and social harm and would almost certainly have saved more lives too. He makes his case with both passion and logic in these pages.

Why did the world go mad? Why did the blunt and brutal policy of lockdown become the one measure that almost all politicians and scientists in almost all countries agreed was unavoidable? After all, quarantining the sick and vulnerable had always been the response to outbreaks of infectious disease, never the quarantining of everybody.

The first reason was surely that for the first time in history we could. Enough commerce had migrated online that if almost everybody stayed at home, much of society could still function – especially the office workers who take decisions in government. The logistics of online retail and the technology of video conferencing had reached some sort of tipping point. Ten years earlier, we would surely not have contemplated lockdown.

A second reason was that China locked down. The attempt to eradicate the virus at source in Wuhan was brutal but largely successful. A policy that was never on the table suddenly became imaginable. There was not a little tinge of envy in some of the early comments made in the West about the power of a totalitarian regime to shut down an entire society. Neil Ferguson of Imperial College put it well: 'It's a communist one-party state, we said. We couldn't get away with it in Europe, we thought . . . and then Italy did it. And we realised we could.' Erstwhile democrats discovered in themselves a surprising love of emergency executive orders.

The reason China tried lockdown was because of SARS. An epidemic had been halted in its tracks in 2003 not just in China but in neighbouring countries and Canada by draconian action. The result was the rapid eradication of the SARS virus altogether. It went extinct except in laboratories (from whence it leaked at least five times over the next year, but that is another story). The Chinese response to Covid, echoed in South Korea and Taiwan in particular, was aimed at suppression of the virus to oblivion. By the time the virus was spreading in ski resorts, hospitals and sports stadiums in the West, that goal was always impossible. Lockdown, designed to eradicate, was not well suited to protecting the vulnerable.

As Mark shows in this book, another reason was that western countries had always prepared for an influenza pandemic, which resulted in assumptions that everybody was vulnerable, that closing schools would help and that social distancing would work well. Covid was dramatically different, singling out the elderly and sparing children almost entirely, but being highly infectious in those showing few symptoms. Yet months later public announcements on the radio were still parroting the nonsense that everybody and anybody was at risk of death. An ultra-precautious mentality took hold, constantly reinforced by the deliberate policy of inducing fear and anxiety in the population to ensure compliance. This led to foolish mistakes like police harassment of hill walkers, and deadly ones like shipping hospital patients to care homes. Expert advice was badly unbalanced, with the malign effects of lockdown itself – on cancer diagnosis, mental health, loneliness and economic ruin – going largely ignored.

Ironically, another reason that lockdown became the weapon of choice was excessive optimism about timing. In March 2020, we were assured lockdown was a brief interruption to save the hospitals from being overwhelmed. Yet it was never realistic that the virus would just go away that fast. Thereafter, we were told that lockdowns would buy time till a vaccine came riding to the rescue, but this was dangerously optimistic about whether a vaccine would be developed at all, let alone fast. In the event, this miracle did occur, though we appeared to have overlooked the fact that rolling out a vaccine would take a long time, even in a very well-prepared country like Britain. But if the cavalry had not appeared, would we really have kept the country locked down for years? This was not a sustainable policy.

The final argument for lockdown was that there was no alternative. Mark demolishes this in devastating and relentless fashion, demonstrating clearly that his proposal to shield the elderly and cocoon the vulnerable by protecting those who came into contact with them, would have worked far better and done far less other harm. After all, this latter policy was what we suddenly discovered and adopted when the vaccine came along: if the elderly were to be prioritised for vaccination, why not for protection as well?

This is a book that shows what we should have done when the UK was confronted with the greatest national crisis of recent decades. Mark's advice would be important even if these were lessons we could only have learned from hindsight. But this isn't hindsight, this is the story of the advice that he was giving at the time as the crisis unfolded.

Matt Ridley

CHAPTER 1

SOUNDING THE ALARM

Early in the New Year of 2020 I was in my office in Edinburgh reading through media accounts of a puzzling respiratory disease – possibly a viral pneumonia – that had surfaced in the city of Wuhan in eastern China. I was concerned and tried to find out more.

On January 7th a helpful journalist sent me a copy of a report written by the Wuhan Municipal Health Committee. The report said that fifty-nine patients were suspected to have caught the mystery disease, seven of whom had become critically ill. The first cases dated back to mid-December 2019 and many were linked to the South China Seafood wholesale market in the north of Wuhan.

The same day the Chinese authorities announced that they had identified the cause of Wuhan pneumonia: it was a coronavirus.

A pandemic begins

As I'd been studying the emergence of new viruses for more than twenty years, I knew what to look out for. New human viruses usually come from animals, and most of them don't spread well between humans. Some coronaviruses can do though, which meant they were high on the list of viruses to worry about.

The fact that there were already fifty-nine cases in a single outbreak told me that this coronavirus probably did spread from person to person, so it was potentially a pandemic virus. If it was, it was probably already too late to stop it. I knew that a respiratory virus could spread around our highly interconnected world and seed a pandemic in a matter of days and I had just learned that this one had been spreading for weeks. There were still plenty of unknowns, but I was now very worried.

On January 12th Chinese scientists published the new virus's genome sequence – its genetic code. The genome sequence confirmed that it was a coronavirus and told us that it is closely related to the SARS coronavirus. This was more bad news. SARS is an extremely dangerous disease that killed more than seven hundred people in 2003. Only a prompt and vigorous international response had averted a full-scale pandemic. The possibility that SARS might one day re-emerge had been a concern ever since. Now we were facing a SARS-like coronavirus with unknown potential.

On January 21st the World Health Organization reported that there had been over two hundred cases and six deaths from the new virus in China, with further cases in Japan, South Korea and Thailand. We call a large but localised outbreak of disease an epidemic, but when an epidemic spreads to multiple countries across a wide region we call it a pandemic. This was not yet officially a pandemic but there could be little doubt that it was going to become one. The virus had already spread beyond China to three other countries and was most likely present in others that we didn't know about yet.

That same day I sent an e-mail to Catherine Calderwood, the Chief Medical Officer (CMO) of Scotland. Even though no cases had yet been reported in the UK, in that e-mail I said that we needed to start preparing for an epidemic that would affect the whole country, Scotland included.

Four days later, new data were published by the World Health Organization that prompted me to write again, with even greater urgency. I explained that – based on these new data – I estimated that the novel virus was capable of infecting more than half the population, tripling the mortality rate and overwhelming our National Health Service (NHS) within two months. I acknowledged that there was a lot of uncertainty but stressed how serious this could turn out to be.

I received a polite reply to my e-mails telling me that everything was under control. It wasn't, as we were all going to find out in the coming weeks.

Warning the Scottish government's most senior medical advisor that we were facing an unprecedented public health emergency wasn't something I did lightly. Before sending those e-mails to Catherine Calderwood I'd consulted with two colleagues who, like me, have been studying epidemics for many years. One was Jeremy Farrar, Director of the Wellcome Trust, and the other was Neil Ferguson, Director of the Centre for Outbreak Analysis at Imperial College, London. The three of us were in complete agreement and Jeremy and Neil were already in touch with the CMO England, Chris Whitty, and the Chief Scientific Advisor (CSA), Patrick Vallance.

In that flurry of communications in January 2020, we set out how we expected the next few months to unfold.

First, the pandemic would be fuelled by mild cases but with significant mortality in vulnerable groups.

Second, there was little prospect of a vaccine against a novel coronavirus becoming widely available in less than twelve months. Third, the prospects for effective therapies were not much better.

Fourth, case isolation, infection control and contact tracing would be crucial, but capacity to deliver them could be overwhelmed if case numbers rose too high. Fifth, social distancing measures such as restrictions on public gatherings and closures of workplaces and schools would then be needed.

Time was of the essence; every day it looked more certain that a pandemic was heading our way. I continued to share information with the CMO Scotland and brought Sheila Rowan – CSA Scotland – and Anne Glover – President of the Royal Society of Edinburgh and former CSA to the European Commission – into the conversation. I was told that the Scottish government was now working 'to address preparedness', though not what that meant in practice. I wasn't convinced. I'd have expected a lot more urgency if the government had been doing the same calculations that I'd been doing.

Three crucial numbers

To get a preliminary idea of how an epidemic could play out we need three numbers.

The first is called the basic reproduction number – this is a measure of how transmissible the infection is and allows us to estimate the fraction of the population who will be infected.

Next is the generation time, the interval between a person getting infected and infecting others – this sets the timescale.

Last is the infection fatality rate – this tells us how many people will die.

These three numbers drove my initial estimates of the potential scale, speed and severity of a novel coronavirus epidemic in the UK. None of these crucial numbers was known precisely in January 2020, but the most uncertain of all was the infection fatality rate. The infection fatality rate is a ratio of deaths to infections. In the early stages of the epidemic in China it was entirely possible that some deaths were being missed. However, there had to be far greater uncertainty about the number of infections. At that time, only clinical cases were being counted, not milder cases or those with few or no symptoms at all. The more of those there were, the more the infection fatality rate would be overestimated.

On January 25th the World Health Organization published an infection fatality rate estimate of almost 5% – meaning that one in twenty of those infected would die. If the true value was anywhere near that figure we were facing a catastrophe. Over the next few months, as surveillance of mild cases improved, the estimate would eventually fall to around 1%. That was bad enough, much higher than influenza which typically has an infection fatality rate less than 0.1%.

All of this told me that this pandemic could be much more severe and much harder to control than the flu pandemic the UK had spent years preparing for. That is why I wasn't convinced by the reassurances I was receiving. I felt we already knew enough to take even stronger action. Giving advice was the easy part; getting anything to happen proved a lot more difficult.

Gearing up

There was a lot more talking than action in the weeks following that first flurry of e-mails.

The UK government's Scientific Advisory Group for Emergencies (SAGE) met on January 22nd. I heard that it was a surprisingly low-key meeting given the gravity of the situation.

I was hurriedly appointed to a SAGE sub-committee called the Scientific Pandemic Influenza Group on Modelling (SPI-M). SPI-M met on January 27th to discuss the need for mathematical modelling of a novel coronavirus epidemic in the UK.

By the end of the month questions about the novel coronavirus had been tabled in both the UK and Scottish parliaments, so the new virus was on our politicians' radar too.

I finally met with the CMO Scotland on February 28th and

at that meeting gave an update on possible scenarios for the UK's novel coronavirus epidemic. The main messages hadn't changed since January, but we now had more data to support them.

The Scottish government subsequently formed its own expert committee, the Scottish Covid-19 Advisory Group, informally known as SAGE-for-Scotland. I was asked to join the group but we didn't hold our first meeting until March 26th, three days after the UK went into lockdown, by which time the course of the epidemic in Scotland and the UK as a whole was pretty much set.

The lack of urgency was troubling. Scientists were warning in mid-January 2020 of an imminent epidemic on a scale we had not seen since Spanish flu more than a hundred years ago. This was an extraordinary event that demanded an extraordinary response and got a very ordinary one.

It was a different story in Taiwan. The Taiwanese government introduced rigorous health checks for arrivals from Wuhan on December 31st 2019, long before most people in the UK had heard about the new virus. By the end of January, Taiwan was screening all international arrivals for signs of infection. At the same time, isolation of cases and their contacts plus quarantining of anyone deemed at high risk was made mandatory and was strictly enforced.

The UK is not Taiwan but, even so, when the alarm was first raised here we could have reacted with the same sense of seriousness and urgency. We didn't. I can think of several reasons why.

For a start, there was a lack of global leadership. In this kind of situation, that is the role of the World Health Organization, but it failed to live up to one of its prime responsibilities. It didn't even declare a Public Health Emergency of International Concern – a precursor to declaring a pandemic – until January

30th. It didn't declare a pandemic until well into March. This badly undermined the case being made for early action in Scotland, the UK or anywhere else.

Another factor was complacency. Since the swine flu pandemic in 2009, the emergence of a new strain of influenza had been on the UK government's national risk register and we had detailed and rehearsed plans for responding to it. It was assumed that those plans would work for the novel coronavirus, but this wasn't flu and an even more vigorous response was required.

There was also a sense of déjà vu. In the early stages of the 2009 swine flu pandemic some scientists had confidently predicted a crisis on a scale far beyond what transpired, largely thanks to an initial overestimate of the infection fatality rate. Policy-makers might reasonably ask whether scientists were crying wolf again.

Those are all plausible explanations for the lack of action, but I think there was something more: sheer disbelief. We were asking officials and politicians to engage with a scenario lifted from a science fiction movie. They simply couldn't take it in.

Looking forward

For those who could take it in, it was already apparent that 2020 was going to be a traumatic year. Many people were going to die and life was going to change for all of us. The policy-makers may have been slow off the mark but at least the scientists did respond during those early weeks.

My research group at the University of Edinburgh was ideally positioned to study this new threat because we do research on the epidemiology of emerging viruses, meaning that we study their origins, distribution and spread. Accordingly, I begged and borrowed the funding to expand and reconfigure my team to work on the novel coronavirus. The same thing was happening in thousands of research groups in universities, government laboratories and industry around the world. Scientists working on diagnostics, drugs and vaccines got to work as soon as the new virus's genome was published in January.

Beyond scientific research, I had a good idea of what the imminent pandemic would ask of me and my team. I'd been deeply involved in the UK's response to previous epidemics: BSE (mad cow disease) in 1996; foot-and-mouth disease in 2001; swine flu in 2009. I was familiar with the sometimes fraught relationship between science and policy. I'd worked with the media on communicating the science to a public anxious to understand what was happening. I'd experienced the pressures that come when the stakes are high and there is heated debate about how best to proceed.

I knew in January 2020 that I was about to go through all that again and more. I knew it would be the same for many of my colleagues in the UK and around the world. I hoped that science would rise to the occasion and that we scientists could make a difference. What I did not expect was that elementary principles of epidemiology – my own subject – would be misunderstood and ignored, that tried-and-trusted approaches to public health would be pushed aside, that so many scientists would abandon their objectivity, or that plain common sense would be a casualty of the crisis.

I did not expect the world to go mad, but it did.

CHAPTER 2

EARLY DAYS

Humanity is plagued by a multitude of different kinds of infection. No-one knew quite how many until my research team spent years trawling through the medical literature counting them. We found reports of infectious diseases caused by hundreds of different kinds of bacteria, of protozoa, of fungi (yes, fungi, but microscopic ones), of parasitic worms, and of viruses. This book is about a single recent addition to that grim catalogue.

Infectious diseases are responsible for a huge burden of death and disease in many parts of the world, but not so much here in the UK. We have a few thousand cases of tuberculosis a year, mainly imported. HIV/AIDS did kill thousands in the 1980s, and is still with us, but mainly affects known risk groups and nowadays can be treated. BSE (mad cow disease) caused a major scare in the 1990s but, thankfully, fewer than two hundred people died. We had only four cases of SARS in 2003.

The UK's deadliest virus is influenza. It kills thousands – sometimes tens of thousands – every year, though most of us think of flu as an inconvenience rather than a threat to our lives. Swine flu in 2009–10 killed fewer than five hundred people.

We hadn't faced anything as serious as novel coronavirus since Spanish flu killed an estimated two hundred thousand in

1918–19. When it came to handling a major infectious disease epidemic, the UK had no track record at all. In 2020, we would have to learn fast.

Learning fast

The new coronavirus that emerged in Wuhan in 2019 was an unknown quantity, but we already knew about other coronaviruses.

Several different coronaviruses infect mammals and birds. Some are a big problem in farm animals, particularly pigs and poultry.

We knew of six other coronaviruses that affect humans. Four of these human coronaviruses cause mild respiratory infections – common colds – but two are associated with more severe disease, MERS and SARS. MERS coronavirus is a problem mainly in the Middle East, where it is found in camels but occasionally spills over into humans causing a respiratory illness that is even more deadly than SARS.

SARS coronavirus and novel coronavirus are close relatives, though the new virus is even more closely related to SARS-like viruses found in bats. The International Committee for Viral Taxonomy – which adjudicates on such matters – considers all of them to be members of the same species. For that reason, the formal name for the novel coronavirus is SARS-CoV-2. I shall continue to call it novel coronavirus (though I'll avoid the acronym nCoV). The disease caused by novel coronavirus was initially called Wuhan pneumonia, but these days it is frowned upon to name a disease after a place so this was changed to coronavirus disease 2019, abbreviated to Covid-19.

Viruses are not, strictly speaking, living organisms. They are strands of nucleic acid – the molecules that make up the genetic code – wrapped in a coat made of protein molecules.

In order to reproduce, a virus has to hijack a living cell – it could be a human cell, although viruses of one kind or another infect every kind of animal and plant – and re-purpose it for producing more viruses. A virus can only do this if it can get into the cell in the first place, which requires that one of the proteins on the surface of the virus – the key – is the right shape to attach to one of the molecules on the surface of the cell – the lock.

Novel coronavirus uses its spike (or S) protein as a key. The lock – more formally referred to as the receptor – was quickly identified as a cell surface protein called ACE-2. The receptor dictates where the virus attacks the body and therefore the kind of illness it causes. ACE-2 is found on cells in many different organs, including the lungs, heart, kidneys and intestines. The SARS coronavirus uses the ACE-2 receptor as well, so it isn't surprising that the novel coronavirus causes a SARS-like illness.

Knowing the symptoms of any infection is central to diagnosing cases and treating patients. Our understanding of which symptoms are the most reliable indicators of a novel coronavirus infection improved during the course of 2020 and the NHS eventually settled on just three – high fever, new continuous cough and loss of sense of taste or smell – though other health agencies also list headaches, fatigue and diarrhoea.

Back in February 2020 there was a lot of debate about whether people could be infected without showing any symptoms at all, so-called asymptomatic infection. For reasons I shall come back to later on, the Chinese authorities were resistant to this idea, but we now know that asymptomatic infections are common. This makes finding cases considerably more difficult and we certainly didn't find them all.

Most people with symptoms recover quickly but some are less fortunate and go on to develop a pneumonia-like illness. Patients admitted to hospital may need oxygen to help their breathing. A minority of cases need intensive care, and some have to be put on mechanical ventilators. Those patients are very ill and many die. For the survivors, the infection can cause significant lung damage and there is a long list of less frequent complications affecting the kidneys, heart and other organs. There can be other long-term consequences of infection too – including what is now called long covid – though this only became apparent several months into the pandemic.

Given that the symptoms of novel coronavirus infection are variable and easily confused with other illnesses we urgently needed a reliable diagnostic test. Within a matter of weeks we had several, an impressive achievement. These tests use a technique called RT-PCR to detect regions of genetic code unique to the virus.

Developing drugs and vaccines was always going to take much longer. There had been a lot of research into drugs and vaccines for the two most dangerous human coronaviruses – MERS and SARS – but without tangible success. This caused some scientists to be pessimistic about any quick breakthrough for novel coronavirus. However, vaccines had been developed for some coronaviruses of livestock, including a cattle virus that is a distant relative of novel coronavirus.

That offered some hope, but we knew from the outset that the first wave of the pandemic would have to be fought without the help of drugs or vaccines. As a first step, we needed to know more about what we were up against.

Timeline of an infection

In the early stages of the pandemic a lot of effort went into characterising the typical course of a novel coronavirus infection.

An infection typically begins when a person breathes in virus particles that then enter ACE-2-expressing cells in the thin layer of tissue – the epithelium – that lines the respiratory tract. Once the infection is established it spreads to other body tissues. For the first day or two levels of virus remain low, but they quickly build up, reaching a peak at around four to six days after infection. At that time the patient may develop symptoms and a few days later the immune system kicks in and levels of virus begin to fall. If severe illness develops it typically does so about a week after the first appearance of symptoms. Most patients that go on to die do so between one and four weeks after falling ill.

This is all important information but we also need to know when a virus infection can be transmitted to other people. Viruses get from one person to another in a number of ways: Ebola transmits through contact with bodily fluids; rotavirus through contact with faecal matter; HIV by sexual contact; Zika virus is picked up by the bite of a blood-feeding mosquito and passed to the next person by the same route.

Respiratory viruses – such as influenza, SARS and the novel coronavirus – are transmitted when we exhale, cough, sneeze or vocalise. For this to happen the virus must be replicating in the upper respiratory tract – the nose, throat and pharynx. Infected cells burst, releasing virus particles and the whole cycle begins again.

Respiratory viruses have always worried epidemiologists on the look-out for pandemic threats because it is so hard to stop them spreading. Transmission happens when people come into close contact, which they do all the time. In the absence of a vaccine, to stop a respiratory virus from spreading you have to stop people behaving as people normally do.

There is surprisingly little data on the transmission of respiratory viruses between humans but – working with Bryan Charleston and his team at the Pirbright Laboratory in Surrey – I have studied the transmission of animal viruses for many years. In early 2020 it just so happened that we were completing some experimental studies of influenza virus transmission in pigs.

Influenza virus is not a coronavirus, and pigs are not people, but I thought our findings might be relevant to novel coronavirus in humans too, if only to suggest what we should be looking out for.

First, even under experimentally controlled conditions, there is a great deal of variability between pigs: some are highly infectious for a prolonged period, others only briefly if at all.

Second, the amount of virus in a nasal swab is a good indicator of how infectious a pig is.

Third, levels of virus decline after a few days and the pigs cease to be infectious, even if we can still detect virus fragments.

Finally, the pigs can transmit the virus perfectly well without showing any symptoms at all.

All these features did turn out to apply to novel coronavirus infections in humans. The last one on my list – transmission without showing symptoms – is particularly important. It was quickly established that even asymptomatic cases had detectable levels of novel coronavirus in the upper respiratory tract, which meant they were likely to be infectious. We now know that asymptomatic cases are about one-third as infectious as symptomatic cases. We also know that symptomatic cases are infectious for one or two days before symptoms appear – the pre-symptomatic phase. This makes sense as we can detect high levels of virus during that phase of the infection.

These are important facts to know. A pandemic will be extremely difficult to control if it is helped on its way by large numbers of people who are infectious but are not yet showing – and may never show – symptoms.

Throughout January and most of February 2020 the majority of novel coronavirus cases were still in China. The Chinese National Health Commission and the China Centre for Disease Control were managing the epidemic but information was slow to emerge, and often did so in the form of official pronouncements rather than scientific reports, which made it hard to evaluate crucial evidence.

Even well into March we did not have a good estimate of the infection fatality rate, though the consensus was beginning to settle on a value of around 1%. We still did not know how many asymptomatic infections there were.

Early pandemic response

Meanwhile, China had taken drastic steps to contain the virus. On January 23rd Wuhan – a megacity of eleven million people – was cut off from the rest of China and put into an extraordinarily strict lockdown. Such an intervention was unprecedented in modern times and public health experts – myself included – were sceptical that it would work. It did and it didn't. It worked in terms of bringing the epidemic under control in Wuhan within a few weeks and slowing the spread across China, but it failed to contain the virus within China.

By the end of February, cases had been reported from forty-eight countries. This did not deter a World Health Organization mission to China from concluding that '*China's bold approach... has changed the course of a rapidly escalating and deadly epidemic*'. It hadn't, but those misleading words were to pave the way for more lockdowns.

It wasn't until March 11th that the World Health Organization declared a pandemic, though epidemiologists had been urging them to do so for weeks. By that date, more than one hundred and twenty thousand cases and four thousand deaths had been reported from over a hundred countries from all around the world. It was patently obvious we were in the midst of a pandemic.

Most countries, including the UK, were already taking action and, for them, the World Health Organization's long-delayed declaration was pretty much irrelevant. The more consequential part of the announcement of the pandemic was the Director General's call for 'urgent and aggressive action' to bring the virus under control, using China as example of success. He did this despite the profound doubts of many public health experts – again including me – that China's lockdown strategy was the best approach to use in the rest of the world.

The UK had reported its first two cases of novel coronavirus on January 31st, both imported from China. By the end of February, the count had increased to twenty-three cases but there had not been any deaths.

The UK government's pandemic response strategy in early March was badged as Contain, Delay, Research, Mitigate.

The Contain phase revolved around using case finding and contact tracing to prevent infection becoming established in the community.

Delay meant slowing the spread in the community using interventions designed to reduce the transmission rate – these could include social distancing measures that reduce the number of contacts between people.

Research covered the development of diagnostics, drugs and vaccines.

Mitigation was about patient care and NHS capacity. Preparing the NHS was a major focus of government planning, culminating in the rapid construction of Nightingale hospitals in England and Louisa Jordan Hospitals in Scotland.

The UK government announced on March 12th that it was moving from the Contain phase to the Delay phase and was abandoning testing in the community so that our still limited testing capacity could be used in hospitals. Though it was controversial even at the time, it was a defensible decision, for two reasons.

First, we couldn't deploy – in the community or anywhere else – testing capacity we hadn't got. That problem had its roots back in January when – had the government taken on board the seriousness of the situation – we could have started to build testing capacity immediately.

Second, it was already too late, novel coronavirus was firmly established in the UK by mid-March and containment was no longer feasible. When the UK government made the switch from Contain to Delay they said that their decisions going forward would be based on careful modelling. So, it's time to talk about models.

CHAPTER 3

MODELS

Scientists use mathematical models to study complex biological processes at every conceivable scale from single molecules to ecosystems. An epidemic has fewer moving parts than an entire ecosystem but it's still extremely difficult to predict how it will play out, even if you're an expert in public health. Models can help.

Epidemic maths

An epidemic is the consequence of infection being transmitted from person to person. The crucial characteristic of this process is that the more infections there are in the population the greater the risk that an uninfected member of that population gets infected. The technical term for this characteristic is positive feedback and it is why epidemics grow exponentially.

Exponential growth is multiplicative (for example, 1, 2, 4, 8, 16, …) rather than linear (1, 2, 3, 4, 5, …) – it will be a recurring theme in this book. An epidemic can't grow exponentially for ever – sooner or later it has to slow down because there are fewer people left to infect – but it can do so in the early stages, as we would soon see with novel coronavirus.

Numbers of cases of non-communicable diseases like stroke or diabetes do not behave this way. For those diseases we sometimes find clusters of cases within families or in a local area but there is no person-to-person spread, so no positive feedback and no exponential growth.

Positive feedback makes it hard to say what will happen if you put in place an intervention designed to reduce the rate of transmission, such as lockdown. If, for example, a lockdown halves the transmission rate (by halving the number of people we come in contact with) it's not obvious what that would do to the size of the epidemic. We'd expect it to be smaller, but would it be half the size, or more, or less? Models are good for answering that kind of question. (The answer, by the way, is that sometimes halving the transmission rate has little impact on epidemic size, sometimes it has a big impact and sometimes it stops the epidemic from taking off altogether – it depends where you're starting from. Epidemics are complicated.)

This is much more than an academic nicety. If those commenting or advising on the response to novel coronavirus do not understand the dynamics of epidemics – and many clearly did not – then their comments or advice can be misleading. Throughout 2020 we saw one example of this after another in public discussions of the R number, herd immunity, elimination, travel bans and the second wave. The resulting confusion clouded those important and necessary debates about what to do next and increased the likelihood of getting crucial decisions wrong.

My favourite example of epidemiological models outperforming expert opinion involves the work of a pioneer in the field, Roy Anderson of Imperial College. Roy and his colleagues modelled the future scale of the HIV/AIDS epidemic in the late 1980s when it was a new and still relatively rare disease. Their predictions of millions of deaths globally were ridiculed by public health 'experts' who failed to grasp that – unlike more familiar infections such as flu – this would be a long, drawn-out epidemic and difficult to stop. As the world now knows, Roy's models were right and the critics were wrong.

SPI-M

For all these reasons, SPI-M – the modelling sub-committee of SAGE – played a pivotal role throughout the UK coronavirus epidemic. SPI-M was co-chaired by Graham Medley – who had invited me to join the committee – and Angela McLean, Deputy CSA. I knew Graham and Angela from when we all worked together in Roy Anderson's team at Imperial College in the 1980s. SPI-M is a large committee and the meetings were always lively. The discussions were well-informed and uninhibited, ideas were put forward and challenged, every session was a crash course in our rapidly expanding knowledge of coronavirus epidemiology.

In happier circumstances, working with SPI-M would have been a joy, but the meetings during 2020 always had an undercurrent of deeply felt concern. The whole point of the modelling was to help us glimpse the future, and none of us liked what we saw there.

Since the models were informing literally life-or-death decisions it was vital that the outputs and the recommendations coming from SPI-M were as robust as possible, and that any uncertainty was reported alongside the headline results. To achieve this, SPI-M uses 'ensemble' modelling.

Ensemble modelling works by consolidating the outputs of multiple, independent models rather than reporting only the output of a single 'best' model. If the models agree, we are more confident in our conclusions. If they disagree, we look for the reasons why and, in the process, may learn something about questions we still need to resolve. Ensemble modelling is a tried-and-trusted approach to modelling complex problems where there are many uncertainties. Climate change modelling works the same way.

SPI-M was a tremendous resource for the UK government in 2020, an assemblage of top scientists in the field, with a structure ready and waiting to deliver modelling outputs in real time as needed, working under the direction of the Department of Health and Social Care (DHSC).

That said, SPI-M frequently found itself at the centre of the storm. The models were (quite rightly) intensely scrutinised and strongly challenged. When, for reasons we'll come to later, trust in the models declined, there was plenty of negative publicity, not least in the November 2020 BBC2 television documentary *Lockdown 1.0 – Following the Science?*

SPI-M had to contend with one challenge right away; it was set up to tackle the wrong disease. That's apparent from its full name: the Scientific Pandemic Influenza Group on Modelling. The committee was created in the wake of the 2009 swine flu pandemic to model the much-anticipated next influenza pandemic (something we may experience in coming years).

For the most part, the influenza expertise in SPI-M was an asset; many of the epidemiological drivers – particularly human behaviour – are similar for novel coronavirus and for flu. Nonetheless, adjustments were needed and some of the models had to be re-written from scratch.

Even then, the new, bespoke coronavirus models betrayed their influenza pedigree. A good example is that the models included schools but not care homes. Schools were the major driver of the 2009 swine flu epidemic in the UK but they were not the major driver of novel coronavirus – this is one of many features of its epidemiology that are more like SARS than flu. Care homes, on the other hand, were crucial. A large proportion of deaths due to novel coronavirus occurred in care homes. As we were to learn, the course of the epidemic in care homes was distinct from that in the wider community – that needed to be modelled explicitly.

Another missing feature was shielding of the vulnerable. Shielding featured prominently in the government's response from March 2020 onwards, but most of the modelling of this intervention was done by independent researchers. Shielding surely had some effect on transmission to people most likely to die – that was what it was intended to do, after all. It was also likely that many people at risk were taking precautions of their own – independent of government advice. The influenza models didn't capture any of this.

A striking omission from the original influenza models was lockdown. A full lockdown was not part of the UK's planning for an influenza pandemic. The models included options for some elements of social distancing – notably travel restrictions and closing schools and workplaces – but not an instruction for most of the population to stay at home. Over the past ten years, SPI-M had built up an evidence base for the interventions we expected to use to control a pandemic. Now we were contemplating doing something different. We'd done our homework, but we'd prepared for the wrong exam.

The influenza legacy was a weakness, but the novel coronavirus models were still useful tools. They were flexible enough to answer the many different questions being asked by government, such as when cases would peak, how much intensive care unit capacity would be needed or what difference more testing would make.

I told a House of Commons Select Committee in November 2020 that decisions on how to respond to an epidemic like novel coronavirus should not be taken without input from models. As I've explained, they provide insights into how the epidemic might unfold that couldn't be obtained in any other way, particularly in the early stages.

That said, I wouldn't want decision-making to be over-reliant on models either. In a Good Practice Guide for modellers I co-authored in 2011 we stressed that models should only be one of the lines of evidence informing policy-makers, never the only one.

In March 2020, however, you could easily get the impression that the UK government's mantra of 'following the science' boiled down to following the models. That's how it looked and that's how the media presented it.

The R number

One of the most prominent outputs of the models was an estimate of the R number. The formal term for R is the 'case reproduction number' and it has a simple definition: R is the average number of cases generated by a single case. The single case is referred to as the index case, and the cases that it generates are known as secondary cases.

The R number is a measure of how well the infection is spreading in the population. R greater than one means that, on average, an index case generates more than one secondary case and the number of cases will grow. R less than one means that, on average, each case generates less than one case and the number of cases will decline. R equals one means that the incidence of new cases stays the same. The phrase 'bringing the epidemic under control' implies reducing the R number from above one to below one.

For a given infection in a given population, the maximum possible value of R is called the basic reproduction number, written as R0 and pronounced 'R nought'. For novel coronavirus in the UK in March 2020 R0 was about three. At that stage of the epidemic, the case reproduction number, R, was equivalent to the basic reproduction number, R0, but once steps were being taken to reduce transmission rates the R number fell and it would fluctuate between 0.7 and 1.5 for the rest of the year.

SPI-M put a lot of effort into those estimates of the value of R. The outputs of these calculations were reported by DHSC as the government's official weekly R number. The R estimate was never a single value, always a range, the size of the range reflecting the degree of uncertainty in the true value. There is inevitably some uncertainty because, in the absence of detailed and systematic tracing data telling us directly how many secondary cases are generated by an index case, we can only measure R indirectly, estimating it as best we can from the epidemiological data available.

Though R is important to infectious disease epidemiologists, it isn't that useful as an operational public health tool.

R doesn't tell you about the number of infections in the population – that's the prevalence – or the number of new infections – that's the incidence.

R only refers to infections, so it doesn't tell you about the numbers of hospitalisations and deaths, which matter most to the NHS and government.

Nor does R help anyone understand their individual risk, which matters most to you and me.

R does tell you about the trajectory of the epidemic but not how fast the epidemic is growing or shrinking. You get that from the doubling time, which I've long argued is a more useful number to communicate than R.

And on top of all that, the methods SPI-M used gave an estimate of R over the past two weeks or so rather than its current value, so it was out of date by the time it was published.

Given these limitations, I was concerned when 'keeping R below one' became a policy objective. SPI-M's weekly estimate of the R number was hugely influential, widely reported in the media and often discussed by politicians and commentators, most of whom had never heard of R before the pandemic began and clearly didn't understand it. As one of my colleagues put it: we've created a monster.

The R monster turned out to be quite dangerous. The policy objective of keeping R below one was often expressed as 'suppressing the virus' and became the central justification for lockdown. The problem is that suppressing the virus is not the only nor the most direct way of minimising the public health burden. The relentless focus on the R number detracted from the usual public health priorities of saving lives and preventing illness.

Herd immunity

Another feature of the models that was widely discussed but poorly understood was herd immunity. Here's how it works. When we are infected by a virus our immune system responds, clearing the infection and creating an immunological memory that protects us against reinfection. Immunity is important to individuals, and it is important to the population as a whole as well. If some people are immune to infection then it is more difficult for a virus to spread through the population, which means the R number will be lower.

The fraction of people that are immune tells us the level of herd immunity. If a large enough fraction is immune then the R number is reduced to one without any additional interventions. That fraction is called the herd immunity threshold.

I've explained what herd immunity is, but we need to be equally clear about what it is not. Herd immunity does not mean that everyone in the population is immune to infection. Nor does it mean that the virus cannot transmit at all in the population (which is why many find the alternative term 'population immunity' confusing, so I will stick with 'herd immunity' in this book). If the herd immunity threshold is passed, a large epidemic is not possible but there may still be outbreaks. The good news is that – because R is less than one – any outbreaks will be self-limiting, meaning that they die out of their own accord. Most of these self-limiting outbreaks will be small, though if R is not much below one they could stretch to hundreds of cases.

The fraction of people that must be immune to achieve herd immunity is related to the basic reproduction number R0, which we can think of as the maximum possible value of R. A crude but useful estimate of the herd immunity threshold is given by the formula: 1 minus 1/R0. If novel coronavirus has an R0 of around three then the formula tells us that the herd immunity threshold is about two-thirds, or 67%. Measles has an R0 of around ten and so the herd immunity threshold is about 90%. Swine flu had an R0 of around 1.5, giving a herd immunity threshold of about 33%.

I must stress – because some commentators were confused on this point – that the herd immunity threshold is *not* an upper limit to the number of people who would be infected during an uncontrolled epidemic. That number – the attack rate – could be considerably higher.

With this in mind we can turn our attention to the heated debate about the role of herd immunity in the UK's epidemic response. In February and March 2020, there was lively argument about the pros and cons of a 'herd immunity strategy' for tackling novel coronavirus. It was a difficult debate to follow for a couple of reasons.

First, the concept of a 'herd immunity strategy' was completely new – the term didn't exist before 2020, so there was no consensus about what it meant.

Second, no-one could argue against herd immunity – that would be irrational, herd immunity works in our favour. Opponents of a herd immunity strategy were really arguing for the strongest possible suppression of the virus. The conflict arises because suppressing the virus has the effect of slowing the build-up of herd immunity by preventing people becoming infected in the first place.

This tension between the short-term benefits of suppressing the virus and the long-term benefits of herd immunity was unavoidable, whatever we did. Epidemiologist John Edmunds put it neatly when he said that 'herd immunity is how this will end, sooner or later'. He meant that the epidemic would continue until enough people had either been infected or vaccinated.

This is why Patrick Vallance and other government advisors – myself included – were talking about herd immunity in March 2020. It was important that we did so. Admittedly, some public pronouncements on the topic were poorly phrased, allowing critics to claim that one of the policies being considered was not to suppress the virus at all. That extreme 'do-nothing' scenario was modelled but – as we shall see shortly – it was never a policy option. Quite the reverse, the do-nothing scenario was modelled to make the case for intervention.

Regardless, Richard Horton – editor of *The Lancet*, a leading medical journal – and others continued to rail against their straw man version of a herd immunity strategy. One of my colleagues was struck by the irony of this, making reference to *The Lancet*'s publication back in 1998 of a notorious paper that raised doubts about the safety of the MMR vaccine. The paper was deeply flawed, and was eventually retracted, but the alarm it caused led to reduced uptake of MMR and the UK almost losing its herd immunity to measles, a public health disaster only narrowly averted.

The debate descended into farce. Some said they disagreed with the idea of a herd immunity strategy because they disap-proved of people being likened to a 'herd' – a nadir for the level of public discussion during the pandemic. On one occasion, one of my colleagues was clearly taken aback when asked on BBC Radio 4's *Today* programme if he 'believed' in herd immunity. That's like being asked if you believe in tides. Like the tide, herd immunity happens whether you believe in it or not.

One of the ironies of this debate was that most epidemiologists thought that herd immunity was almost irrelevant in March 2020. That's because too few people had been infected – and so might now be immune – to make any discernible difference to the trajectory of the epidemic.

There was a minority view that this was wrong, that many more people had been infected than we thought because the virus had been circulating for longer than we were assuming. This idea was fuelled by a trickle of anecdotal reports from people who claimed to have had Covid-19-like symptoms in January 2020 or even earlier.

Serological surveys – which test for antibodies against the virus and so indicate how many people have had the infection – eventually put paid to this theory. Data published from April 2020 onwards indicated that no more than 10% of the population had been infected in London, and only around 5% in other parts of the UK. This tallied with expectations from my team's mathematical models and didn't support the idea of large, hidden epidemic.

The UK never adopted a herd immunity strategy, but the debate left a legacy of misunderstanding. Months later, I was taken aback when some research done by my team was criticised as being based on a 'herd immunity model'. Of course our model included a representation of herd immunity. That didn't make it controversial, it made it a standard, text book epidemiological model of a human viral infection. All the novel coronavirus models incorporated herd immunity – they would be deficient if they didn't.

Using models to make predictions

Epidemiological models have various uses. They are often used to interpret epidemiological data; for example, helping to assess in retrospect the impact of interventions. They can also be used to make predictions.

Prediction was challenging for novel coronavirus in the first half of March 2020. Accurate prediction of an epidemic is hard enough even for infections we know a lot about, and there were still many gaps in our knowledge of this new virus. The models were not data-free – there was information from China and a few other countries – but we were at too early a stage of the UK epidemic to calibrate the models against UK data. That left plenty of scope for uncertainty.

The role of herd immunity was one area of uncertainty. I said earlier that herd immunity is an integral feature of virus epidemiology, but there's a complication. For some infections – including some coronaviruses – immunity is not 100% protective and wanes over time. Until we knew whether either of these applied to novel coronavirus we couldn't make any useful predictions of the long-term course of the epidemic. Fortunately, this wasn't such a problem for short-term predictions – as we saw earlier, herd immunity didn't play a significant role in the early stages of the epidemic.

The biggest challenge to making short-term predictions was the role of human behaviour. The spread of any respiratory virus depends on how people behave, especially the number and nature of the contacts they make with other people.

We know a lot about people's contact behaviour in normal circumstances. As part of the preparations for pandemic influenza there had been several large-scale studies of volunteers recording who met with whom and how often. These data were an essential input into the epidemiological models. However, we were now expecting those behaviours to change; that was the whole point of social distancing measures.

The problem was that we didn't know exactly *how* those behaviours would change. That would depend on how well people complied with advice and regulations, on what they did instead of doing the things they'd been told not to do, and on how they changed their behaviour of their own accord, independently of government directives. Basically, we were being asked to predict how people would respond to a once-in-a-lifetime crisis and there was no sound basis for doing so. The best we could do was guess.

If I had to pick anyone to guess how people's behaviour might change during a pandemic then I would choose a group of scientists who have been studying this kind of problem for many years – I would choose SPI-M. At the same time, it's important not to overstate how much confidence we should have in those guesses. The best we can do is try out different assumptions – assumption is a more scientific-sounding term for guess – about how people might behave and proceed from there.

How people will behave is not the only uncertain aspect of the epidemic. We might also want to explore different assumptions about immunity, the transmission rate, the generation time, the infection fatality rate and every other component of our model. On top of that, we will often want to compare the expected effects of different interventions or mix of interventions.

I shall use to the word 'scenario' to refer to a single combination of assumptions and modelled interventions. A scenario is a list of 'if' statements: if the virus does this... and if people behave like this... and if the government does this... then we expect...

You will immediately see that at any point in the epidemic there could be lots of possible future scenarios to consider.

With modern computers that's not a problem – we can look at a huge number of scenarios extremely quickly – my team sometimes looked at a million in a single modelling study. Even then, we didn't necessarily expect any of the million to turn out to be precisely correct. We were much more interested in what the exercise told us about the range of possible outcomes. Where there was consistency we could be reasonably confident that we knew (at least roughly) what was likely to happen. On the other hand, if there was a wide range of possible outcomes and we had no good basis for preferring some scenarios over others then the wisest course was to say that we don't know what will happen and wait for more data.

Use and misuse of the reasonable worst case scenario

Mind you, even if the future is uncertain, governments and health care providers still have to plan for it. One useful planning tool is a scenario called the reasonable worst case. This scenario isn't the most likely outcome – it's at the extreme end of the plausible range where cases, hospitalisations and deaths are highest – but you can't categorically rule it out. The reasonable worst case boils down to 'hope for the best but plan for the worst'. If you plan for it – for example, in terms of building extra hospital capacity to deal with a possible surge in severe cases – then you shouldn't get caught out.

One problem with the reasonable worst case scenario is that the media cannot resist treating it as a prediction. There was an outcry in 2019 when 'Operation Yellowhammer' was made public. This was a planning exercise based on a particularly gloomy scenario for the state of the UK in the weeks following a 'no-deal' departure from the European Union. It was treated by the media as though it were a prediction and got plenty of attention. The reasonable worst case is not a prediction – we'd be (unpleasantly) surprised if it happened – it's a planning tool.

That brings us to Imperial College's Covid-19 Response Team Report 9, published on March 16th 2020. The model used to produce this report generated a worst case scenario of over half a million deaths due to novel coronavirus in the UK by the end of July. This wasn't exceptional: all the models suggested that novel coronavirus was capable of causing a huge epidemic. I'd shared such a scenario with the Scottish CMO back in January. The problem was that these worst case scenarios weren't realistic and weren't intended to be.

The worst case epidemic is one in which the virus is allowed to spread in the absence of any countermeasures at all. This is the do-nothing scenario we encountered earlier. You might call it the *un*reasonable worst case scenario. It's useful to know how big that epidemic would be but it would never happen in practice. It is inconceivable that we would all carry on as normal while hundreds of thousands of people were dying, even in the purely hypothetical event that government did nothing at all.

Report 9 went on to look at the effects of different combinations of interventions – self-isolation of cases and household contacts, school and university closures, and full lockdown – over a period of two years. I was extremely sceptical about this. It was perfectly obvious that no-one could predict the course of this epidemic over such a long timescale, so what was the point of publishing these outputs?

What's more, the outputs the study wasere based upon a single set of assumptions – or, if you prefer, guesses – about the impact of each intervention. That was a concern because no-one could be sure what those impacts would be in practice, introducing massive uncertainty into the analysis.

Even if Report 9 wasn't a good justification for a full lockdown, it did contain one important result: if we tried to tackle the novel coronavirus epidemic mainly by social distancing then we were in for a torrid time. The report should have been a wake-up call that we needed to invest quickly and heavily in other ways to control novel coronavirus or – according to the model – we'd end up in lockdown. This implication was barely mentioned – lockdown was accepted as a necessity the first time it was proposed.

When Report 9 was published the details of the scenarios modelled were quickly forgotten, as were any mentions of the assumptions, caveats and uncertainties of the analysis. Report 9 was condensed to the simple but misleading message that, if the government didn't impose full lockdown immediately, over half a million people would die.

My last on-camera TV interview before lockdown was an attempt to alleviate some of the huge public anxiety created by Report 9. The BBC interviewer seemed genuinely rattled by the frightening 'prediction' of so many deaths. I tried my best to explain that this was the kind of exercise that mathematical modellers undertook to understand the epidemic better, it wasn't realistic and it wasn't a prediction of what was going to happen come what may. I don't think I succeeded.

Defining the strategic objective

Half a million deaths may have been implausible but we were still facing a public health emergency. The challenge facing SPI-M in February and March was to work out how best to meet the UK government's stated policy objectives: to save lives and protect the NHS. This wasn't as straightforward as it might sound.

Let's start with the first objective, saving lives. If that objective is interpreted as trying to minimise deaths due to novel coronavirus while ignoring deaths from other causes, and if social distancing is the intervention of choice, then we don't really need a complex computer model to tell us what to do. The best solution is to lock down immediately, as tightly as possible, and wait until some alternative – such as a vaccine or effective treatment – presents itself, regardless of how long that might take. There was no shortage of people advocating this approach, but none of the UK administrations pushed the 'saving lives' objective to its logical conclusion.

The second objective was to prevent the NHS from being overwhelmed. This implies that an R number above one might be manageable as long as the resulting epidemic was not so large that the NHS couldn't cope, which was usually interpreted as not running out of intensive care capacity. Prime Minister Boris Johnson characterised this strategy as 'flattening the curve' and 'squashing the sombrero'. Unfortunately, there were two problems with the approach.

First, we come up against one of the paradoxes of the dynamics of epidemics. As you'd expect, a higher R number results in a larger epidemic with a higher peak number of cases and, therefore, a higher peak number of hospitalisations and deaths. An R number that is lower (but still above one) results in a smaller epidemic with a lower peak. That seems straight-forward enough. Unfortunately, the smaller epidemic is also more prolonged. That's not obvious at all – as we saw at the beginning of this chapter, intuition is often a poor guide to how an epidemic will play out. If the epidemic was going to be a long-drawn-out affair then we needed to think not only about the NHS's surge capacity but also about sustainability.

That brings us to the second problem: the NHS does not have a great deal of spare capacity at the best of times. Whether there was a huge surge in demand for a limited period or a more modest increase spread out over many months, the NHS couldn't cope with either.

This meant that flattening the curve was only ever a realistic strategy if it were linked to the provision of both additional infrastructure – more beds and more intensive care units – and many more trained staff. We got the infrastructure – in the form of the Nightingale and Louisa Jordan hospitals – but staff numbers remained limiting.

Looking back, it seems perverse that the UK tried to combat a major public health emergency without making a more determined effort to boost health service capacity. This was one of many instances where we were handicapped by a failure to appreciate the seriousness of the situation we were facing and a reluctance to accept that the crisis would not be over in a few weeks.

By default, SPI-M was left with trying to identify a strategy which was more than flattening the curve but less than all out suppression – not an easy task. I recall a heated discussion of whether it was possible to maintain the R number at round about one. I described this as 'utterly unrealistic' on the grounds that we had no way of knowing what combination of countermeasures would be required. This was when I first realised that we couldn't try to control this epidemic simply by managing the rate of transmission. There was too little room for manoeuvre. We'd end up yo-yoing between intolerably severe restrictions and unsustainable pressure on the NHS (and that, of course, is exactly what happened).

My team started looking for alternatives right away, but it was to take a couple of months before we had a firm proposal to bring to the table. In the meantime, there was a good argument that it was better to err on the side of caution – if the NHS were truly overwhelmed the consequences would be horren-dous, and not only for patients with Covid-19. I fully agreed with this – I had highlighted the danger of the NHS being overwhelmed back in January. The problem was the proposed solution: lockdown.

CHAPTER 4

LOCKDOWN

The UK's novel coronavirus epidemic took off in March 2020. By the middle of the month the daily count of new cases was over seven hundred and the daily death toll was more than forty. Even more worryingly, both numbers were doubling every three to four days. The epidemic was out of control and we were playing out the scenario that had been put to the Chief Medical Officers of England and Scotland in January.

Why a rapid response was vital

Not everyone understood quite how quickly we were approaching a crisis point, however, because not everyone understands exponential growth. There's a classic biology exam question that goes like this: if the number of bacteria growing in a flask is doubling every half hour, and after twenty-four hours the flask is full, when was the flask half full? The answer is obvious but striking: just half an hour ago. For twenty-three of the twenty-four hours you might not be the least bit worried about the flask overflowing, but that changes very quickly indeed.

The early stages of an epidemic work the same way. If the epidemic is doubling in size every three or four days then a

Covid-19 ward that had been open for weeks but was still only half full one Friday morning could be overflowing by Monday afternoon.

In real life, the situation was even more precarious than that because the trajectory of the epidemic cannot be changed instantly, it has momentum. This is because it typically takes around a week for a newly infected person to develop symptoms and another week to land in hospital if they are unfortunate enough to fall seriously ill. Whatever trajectory hospital admissions are going to take in the next two weeks, nothing can be done to alter it now. By March 23rd SPI-M modellers were predicting that there was no time to lose if we were to prevent the NHS being overwhelmed in early April.

There has been plenty of discussion about the role that the models played in the decision to go into full lockdown on March 23rd. Imperial College's Report 9 of a do-nothing scenario resulting in over half a million deaths is said to have been particularly influential, even though that scenario wasn't remotely realistic. For me, it wasn't the models and their speculative long-term predictions that decided the issue, it was the short 'doubling times' of cases and deaths and the inevitability of the situation getting considerably worse before it got better.

At the SPI-M meeting on the morning of March 23rd I supported the committee's recommendation of an immediate national lockdown. It was a sombre moment. I was extremely concerned about the damage that a lockdown would do, but a decision had to be made and there was no other option on the table. SPI-M's advice was echoed elsewhere, was endorsed by SAGE later the same day, and the Prime Minister announced the lockdown that evening. The UK was given a clear and simple instruction: stay at home.

Immediate impact of lockdown

As we'd expected, the epidemic continued to grow for two or three weeks after lockdown was imposed, with deaths peaking at over a thousand a day during the second week of April. It did not go into sustained decline until the end of that month. In the worst-affected regions of the UK, intensive care unit capacity was stretched close to its limit and surge capacity was called into play, though the rapidly constructed Nightingale hospitals were barely used.

Before we go into the numbers in more detail, I must acknowledge that the death toll would have been far greater but for the tremendous efforts made by NHS staff during that period. Our hospitals may not have been overwhelmed during the first wave – as many feared they would be – but the NHS still had to deal with more than a hundred thousand patients, many of whom owe their lives to the care they received.

Cases, hospitalisation and deaths declined steadily through May and June, reaching low levels during the summer even as restrictions started to be eased. It seemed that the lockdown had worked. However, if we look more carefully it isn't as straightforward as that. When the UK went into its first lockdown fewer than four hundred people had died and when we came out that number had risen to over forty thousand, which doesn't look like an unqualified success. Many of the infections that led to those forty thousand deaths were acquired in the weeks before lockdown – that's a consequence of what I referred to earlier as the epidemic's 'momentum'. Lockdown came too late to prevent those infections from happening, even though the patients may have died up to four or five weeks after it was imposed.

We might then ask how many deaths were caused by infections that happened *after* lockdown was imposed. I co-wrote a report on this issue with my colleague Chris Robertson in

October 2020. Our answer was that more than half of the fatal infections during Scotland's first wave were acquired after the lockdown began. Much the same is likely to be true for the rest of the UK. Our conclusion from this exercise was that lockdown may or may not have been necessary to protect public health – we'll come to that shortly – but it certainly wasn't sufficient. At best, lockdown was only part of the solution; we needed to do more.

A more difficult and controversial problem is working out how many lives lockdown did save. To do that we need to infer what would have happened if we hadn't gone into lockdown. The 'what would have happened?' question is more formally known as the counterfactual scenario. Choosing the right counterfactual is critical. One theoretical possibility is that if we don't go into lockdown we do nothing, we carry on with business as usual as the epidemic rages around us. As I explained in Chapter 3, it is useful to model that scenario but anyone can see that it's not remotely realistic.

In reality, the UK government introduced a number of measures to try to control the epidemic before we went into full lockdown.

From March 12th people with symptoms were required to self-isolate.

On March 16th advice was issued to avoid non-essential travel, work from home, not go to some social venues and for vulnerable people to shield. On March 18th schools were closed for the great majority of children.

On March 20th most public venues were closed. These measures surely had a significant impact on transmission rates.

In addition to interventions by government, many people were taking their own precautions before lockdown, as were many businesses and institutions. My own university shut down most routine activities well before March 23rd. These voluntary changes in behaviour – not imposed by government – surely had some impact too, but the models did not account for them.

After lockdown was imposed a team at the London School of Hygiene and Tropical Medicine set up a study called Comix to monitor the number of contacts people were actually making week by week. Comix provided invaluable data during the pandemic, but unfortunately it started too late to tell us what was happening during the period immediately pre-lockdown.

The best we can do instead is look at mobile phone data provided by Apple and Google on patterns in people's movements (aggregated so that no information can be traced back to any individual). These 'mobility' data show that there were marked shifts in people's activity patterns in the week *before* March 23rd, with many more people apparently staying at home. Lockdown itself seems to have come late to the party and had surprisingly little effect. This is not good news for anyone wishing to argue that lockdown had a decisive impact on the course of the UK epidemic.

Undaunted, the Imperial College team published a counter-factual analysis of the impact of lockdown in June in *Nature*, a leading science journal. Their analysis allowed an impact of measures introduced before March 23rd but ignored voluntary behaviour changes and assumed there would have been no further changes in behaviour had we not gone into lockdown. That's a kind of do-nothing scenario and – as with the version we encountered in Chapter 3 – it exaggerates the impact of lockdown.

The study concluded that the measures introduced before March 23rd did have some impact but were insufficient to reduce R below one; full lockdown was needed to bring the UK epidemic under control. They drew the same conclusion for several other European countries.

The analysis stumbled badly, however, over Sweden. Sweden never went into full lockdown but it did introduce a suite of countermeasures and it did bring the epidemic under control. The Imperial College model had no choice but to attribute the decisive impact to the final measure implemented, which happened to be the banning of mass gatherings. So we were asked to believe that banning mass gatherings in Sweden had much the same effect as full lockdowns did in other countries where banning mass gatherings had only a small impact.

This conclusion wasn't remotely plausible and the *Nature* paper attracted a lot of criticism (as you can see from the Comments section on the journal's web site). In my view, the *Nature* paper reveals a fundamental weakness of this approach to epidemiological modelling. It assumes that the only drivers of behavioural change during the pandemic are the restrictions being modelled, which can only exaggerate their importance. The conclusion I draw from this exercise is that this crucial assumption isn't consistent with the data and a re-think is called for.

Analyses by other researchers came up with the quite different conclusion that the UK epidemic was already in decline before lockdown took effect, even if it was not possible to know that at the time, because of the time lag of a week or so between infection and the reporting of symptoms. According to this interpretation, the measures put in place before March 23rd, coupled with voluntary behaviour changes, had already reduced R below one. Lockdown happened because by that stage no-one – including me – was prepared to risk waiting to find out if those earlier measures had worked.

I expect this debate will rumble on for years, but doubt anyone would claim now that the March 23rd lockdown saved anywhere near half a million lives. That figure comes from the totally unrealistic counterfactual scenario in which no-one does nothing at all to control the epidemic, and even then it's subject to a huge amount of uncertainty. There were many other interventions available to us short of full lockdown. Anyone who supported lockdown on the basis of the half a million figure was misled.

Getting the timing right

Another hotly contested issue is whether the UK lockdown was implemented too late. Although this debate has been a lot louder after the event than it was before, it became a rallying cry for critics of government policy and epidemiological modelling alike, so it's worth looking closely at the arguments, while trying to distinguish hindsight from what was being said at the time.

I'll begin this discussion by telling you what the conclusion will be. Yes, the UK should have acted earlier, but that action did not have to be a full, March 23rd-style lockdown. At this point, students of infectious disease epidemiology are objecting that they were taught (possibly by me) that epidemic control is all about intervening early and intervening hard, so how can I say that we might not have needed a full lockdown? Let me explain.

The rationale for the hit-early-hit-hard rule is fairly straight-forward. If an epidemic is growing exponentially then there is a disproportionate benefit to intervening sooner rather than later. In the same vein, it's always better to over-react than underreact and suffer exponentially growing consequences of failing to bring the epidemic under control.

Those are pretty good rules of thumb and many basic epidemiology courses leave it there. If we look a little harder though, we find that we have oversimplified the problem. There are hidden assumptions underpinning the hit-early-hit-hard rule that most students never get to learn about.

The first assumption is that the intervention is cost-free or, at least, that the cost is trivial when put against the benefit to public health and so we can reasonably ignore it. It was obvious that we couldn't make that assumption about lockdown; anyone could see that lockdown was going to be hugely damaging in many different ways.

The second hidden assumption is that the intervention can be sustained as long as necessary, perhaps indefinitely. We can't make that assumption about lockdown either, for the same reason.

In February and early March 2020 my team and others in SPI-M modelled many different versions of lockdown: different timings, different severities, different durations, on-off lockdowns and so on. This exercise demonstrated that the dictum hit-early-hit-hard is indeed too simplistic for such a drastic and unsustainable intervention as lockdown. Instead, there is a trade-off between intervening early and intervening hard; the earlier measures are introduced, the less severe they need to be to prevent the NHS from being overwhelmed.

This was a useful result and it is the basis for my claim that by acting earlier the UK could have avoided full lockdown. Even if we had erred on the side of caution and imposed an early full lockdown anyway, we could have lifted some restrictions much sooner than we did without risking a public health disaster.

The prospect of lifting restrictions brings us to another problem with the March 23rd lockdown: the lack of an exit strategy. This was a big concern of mine in February and early March. The only concrete suggestion for an exit strategy was locking down until we had a vaccine. I didn't think that waiting for a vaccine was a strategy, it was more of a hope. What would happen if we didn't get a vaccine for a year, several years, or never, all perfectly plausible possibilities at the time? I never got an answer to that question.

When the UK did go into lockdown, no-one could say with any confidence when we could expect to come out of it. There were glib promises that lockdown would only be for a few weeks, which look astonishingly naïve now.

Now that we have covered the background, it is time to talk about specific dates for an earlier intervention in the UK. In principle, that could have been very early indeed, even before we detected our first cases at the end of January. Some countries – such as Taiwan – had introduced strict border controls by that time. The UK had not, which is hardly surprising as we would have been going against the World Health Organization advice if we had. I'll return to the puzzle of why the World Health Organization was not recommending border controls and travel bans later in this book.

A pivotal date was February 28th. This was the day that the UK reported its first confirmed cases of community transmission, meaning that novel coronavirus was no longer confined to travellers, it was spreading within the UK population. This is significant because, even as far back as January, we were confident that the R number for novel coronavirus in the UK would be considerably greater than one, which meant that once community transmission got under way an epidemic was inevitable. The community transmission criterion was used as a trigger for lockdown by both New Zealand and Australia later in the pandemic when they were trying to protect their coronavirus-free status.

My best estimate is that February 28th was the latest date when the UK could have taken a New Zealand-like route of eliminating the virus and closing our borders. It may have been too late even then. That's useful for future reference, but I'm not aware of anyone suggesting it at the time.

On March 12th Chris Whitty announced the move from the Contain phase of the UK coronavirus response to Delay and, at the same time, the abandonment of community testing for infections. As Anne Glover pointed out at a Royal Society of Edinburgh meeting a few months later, this was a warning sign that the epidemic was out of control. By that date the UK had over five hundred cases and numbers were rising. Nine patients had died. Even so, there was still no concerted call for immediate action. SAGE was 'considering' additional measures.

The pace accelerated over the next few days. Spain followed Italy into national lockdown on March 14th. On March 16th SAGE recommended implementing additional measures as soon as possible. The public discussion in the UK through the week of March 16th was mostly about locking down London, not the whole country. London was about a week ahead of the rest of the UK on the epidemic curve. This was expected because London is by far the UK's biggest travel hub, so the epidemic was always likely to be seeded there first. Even if we'd just locked down London on March 16th that would have made an appreciable difference, perhaps enough to dampen the recriminations that followed.

It's easy to say what I just said: that locking down earlier – whether for London or for the whole country – would have saved lives. As I've explained, you can take that argument as far back as the end of February if you like. I am more interested in whether intervening earlier could have saved lives *and* kept us out of full lockdown. Both of these possibilities are counter-factual scenarios.

As we saw earlier, it's difficult enough to use counterfactuals to estimate the impact of the lockdown we did implement on March 23rd. That requires us to compare what actually happened with a hypothetical alternative. It's even harder – and even more speculative – to estimate the impacts of measures we might have taken on other dates. That's comparing one hypothetical alternative with another hypothetical alternative. This hasn't stopped some modellers from trying but I don't give any of these studies much credence – they use the same kind of flawed approach that I critiqued earlier in this chapter.

It's more instructive to look at what the models were saying before the event rather than after it. The first time my team suggested an immediate intervention to slow the spread of novel coronavirus was in a briefing we wrote for SPI-M on March 4th. This came from the study I mentioned earlier showing that – as a general rule – an earlier intervention could be a less drastic intervention. I mustn't claim too much; the immediate imposition of limited restrictions was only one of several scenarios we presented in the briefing, though with hindsight it was probably the best one.

What might we have done on March 4th? You will recall that later the same month the UK introduced a number of measures prior to going into full lockdown, including self-isolation of cases and restrictions on travel, work, social venues and public gatherings. These were substantial interventions, but we didn't wait to find out if they worked. If we'd implemented those measures on March 4th then we could have waited. I cannot say for sure whether that would have been enough to keep us out of full lockdown, but it would have given us a much better chance.

I do not agree with those SAGE members who have made public their regret that they did not recommend lockdown a week or so earlier than March 23rd. As far as I'm concerned, going into lockdown was a failure in itself – we needed to act even sooner to avoid having to take that step. After all, whenever we did it, going into lockdown was merely swapping one problem for another; we were jumping out of the frying pan into the fire.

A brief history of lockdown

You won't find the word 'lockdown' in a text book on public health published before 2020. Collins Dictionary – who made 'lockdown' one their 'words of the year' for 2020 – define it as the imposition of stringent restrictions on travel, social interactions and access to public spaces.

The package of measures imposed across the UK on March 23rd 2020 is described as a lockdown. So too are the measures imposed on November 5th 2020 and January 4th 2021. Local lockdown of a city or a region is also possible. For the rest of this book I shall use lockdown to refer to a package of legally enforced measures that limit the reasons people can leave their own homes.

Lockdown is an extreme response and – in modern times at least – an unprecedented one. It was bound to cause immense harm and, as we'll see later, it did. The benefits were uncertain too. The immediate goal for the UK in March 2020 was to bring the epidemic under control but it wasn't clear what we would do once that was achieved. If we simply came out of lockdown the number of cases would start to rise again and sooner or later we'd be back where we started.

Lockdown was never going to solve the novel coronavirus problem, it just deferred it to another day, and it did so at great cost. Epidemiologists and modellers knew that was going to be the case from the outset. It turned out that policy-makers did not, but this only became apparent several months later. This is a crucial point; everyone needs to understand what such a harmful intervention can and cannot achieve before we introduce it. They didn't.

One thing lockdown could do, though, was buy us some time. Given the harm that lockdown would do every single day it was in place, it was vital that we used that time well. Nicola

Sturgeon, First Minister of Scotland, articulated this when she spoke of a 'contract' between the Scottish Government – who had imposed lockdown – and the people – who would suffer the consequences.

Back in March 2020, potentially game-changing innovations such as mass testing and vaccines were still months or possibly years away, so the government's side of the bargain wasn't easy to fulfil. In my view, the best any government could do was to find a way to get us out of lockdown as quickly and as safely as possible. I saw lockdown as nothing more than a stop-gap measure justified only by the want of anything better, to be endured for the shortest possible time.

Marc Lipsitch, an epidemiologist at Harvard University, put this beautifully when he likened lockdown to a flimsy life-raft you are clinging to in a storm-tossed sea. It's true that the raft is saving your life for the moment, but it's not a viable long-term option and you need to find your way to dry land as soon as possible. Marc was right; we needed an exit strategy, and we needed it quickly. We never got it; governments around the world clung to lockdown in one form or another throughout the remainder of 2020 and into 2021.

It's helpful to understand why anyone thought lockdown was an appropriate response to novel coronavirus in the first place. As I explained in an article in the *New Statesman* in July 2020, the key to the puzzle is that lockdown was never supposed to be a sustainable public health intervention.

Lockdown was conceived by the World Health Organization and China as a means of eradicating novel coronavirus once and for all from the face of the earth. With hindsight, this plan was doomed from the outset, but the World Health Organization believed they had history on their side. After all, they had been there before. In 2003 a new coronavirus had emerged from China, SARS. The subsequent eradication of SARS within a year is rightly regarded as one of the World Health Organization's greatest triumphs.

The novel coronavirus that emerged in 2019 is a close relative of SARS and the plan was to eradicate it too. This time, despite a massive effort involving the strictest possible lockdown of the entire city of Wuhan, the plan failed.

Part of the reason the eradication strategy failed is because it relied on active case finding. That worked well for SARS because there was little transmission in the absence of symptoms, so isolating symptomatic cases stopped most of the spread. The same isn't true for novel coronavirus but – because it undermined their eradication strategy – China wouldn't accept that there was a significant number of asymptomatic cases.

The failure to eradicate novel coronavirus was disappointing, if unsurprising given that this new coronavirus had spread more quickly, more widely and more surreptitiously than SARS. However, it was the World Health Organization's unwillingness to accept the reality of the situation that set the scene for what was to follow.

One consequence was a stubborn reluctance to admit that we were experiencing a pandemic. This would be an admission that the eradication strategy had failed, even when this was patently obvious to everyone else. The World Health Organization finally gave in and declared a pandemic on March 11th. I am convinced that they should have abandoned lockdown as the primary means of tackling novel coronavirus at the same time. They did not; they praised China's efforts and promoted lockdown. The world was given an intervention that only made sense in the context of eradication as the preferred means to control a disease that was clearly here to stay.

The World Health Organization did eventually start to back away from advocating lockdowns, culminating in a statement by David Nabarro, the World Health Organization Covid-19 Envoy, on October 8th 2020 saying 'We really do appeal to all world leaders: stop using lockdown as your primary control method'. Tragically, this appeal came seven months too late and by that time a colossal amount of damage had already been done.

CHAPTER 5

BALANCING HARMS

During the pandemic several politicians, Nicola Sturgeon for one, adopted the position that 'no death from coronavirus is acceptable'. This isn't a surprising thing for a politician to say – it sounds compassionate, reasonable and unarguable – but political rhetoric should not be confused with practical health care policy. Taken literally, the idea that no coronavirus-related death is acceptable makes it impossible to tackle the novel coronavirus epidemic in a rational manner. Unfortunately, it was taken literally, and not only in Scotland, and that's a large part of the reason why we ended up in lockdown.

Deciding what's acceptable
Public health is always about finding a balance between costs and benefits. Whatever way you dress it up, that always comes down to putting a monetary value on health and life. This is a concept that many people find disquieting but is impossible to avoid as long as health budgets are limited.

Within the NHS the job of finding the right balance between costs and benefits is given to a body called the National Institute of Health Care, or NICE. NICE makes difficult decisions about which medicines, treatments and procedures the

NHS can and cannot afford. To do this, NICE generally uses a valuation of around £25,000 per year of healthy life. Whether that figure is applicable during a pandemic is debatable. Some economists argued that we were in a situation akin to fighting a war, so different rules apply. That may be so, but it does not mean that we can disregard the cost altogether.

For novel coronavirus, the costs of the public health response are not just monetary. We need to find a balance between the illnesses and deaths due to the virus and the multiple harms caused by locking down much of society. The potential for harm on both sides was so great that striking the right balance was always going to be critical for minimising the overall damage caused by the epidemic and its aftermath. The problem the rhetoric had created was that if no death due to novel coronavirus was 'acceptable' then no balance could be struck.

We do not treat any other public health issue in this way. A good example is road safety. One of the biggest contributors to fatal traffic accidents is speed. If we took the stance that 'no death from a road traffic accident is acceptable' then – if we allowed driving at all – we would impose five-miles-per-hour speed limits on every road. If you think that proposal absurd then you are admitting that we are not willing to do everything we conceivably could to reduce the number of deaths on our roads to zero.

We do not treat other infectious diseases that way either. In most winters, thousands of people – mostly elderly – die from influenza across the UK. In 2013–14 it was more than thirty thousand. We do try to prevent these deaths – every year flu vaccinations are offered to around thirty million people – but, if we chose, we could do more. If we went into lockdown every winter that would stop flu from spreading and save thousands of lives. That's because the maximum R number for flu is about half that of novel coronavirus, but they spread in similar ways, so if we can stop novel coronavirus we can certainly stop flu.

This is no mere debating point. The idea that no death from novel coronavirus is acceptable – unlike deaths from road traffic accidents, influenza or anything else – led to some of the worst consequences of the epidemic.

It instigated the collapse in health care provision during the first wave, which was inevitable when the NHS was being asked to prioritise novel coronavirus above everything else.

It devalued the psychological and mental health harms of imposing severe social distancing rules on an entire population.

It legitimised the closing of schools even though – as we shall see later – this had limited public health benefit, and almost none to the children.

It ignored the long-term impacts – not least to future NHS funding – of the economic damage caused by lockdown.

It justified the shutting down of society on the grounds that nothing was more important than preventing deaths due to novel coronavirus.

Harms caused by the virus

Make no mistake, for millions of people, Covid-19 is a dangerous disease. It is not just a 'bad flu', as some have claimed; it is a far more serious threat to public health than that. To understand just how serious, we need some numbers. The public health burden caused by any disease can be quantified in several different ways: counting deaths, calculating the burden on the NHS, and estimating the burden of illness across the entire population.

Counting deaths is an obvious measure, though even that isn't as straightforward as you might think. Attributing the cause of a death is often difficult; many people, especially the elderly, have underlying health conditions and die with multiple causes mentioned on their death certificate. If you add up every cause of death mentioned on every death certificate you would conclude that many of us die more than once.

There's also the problem of distinguishing people who died *of* novel coronavirus infection from those who died *with* a novel coronavirus infection. The Radio 4 programme *More or Less* played with this issue by trying to estimate the number of people hit by a bus who had tested positive for novel coronavirus in the previous twenty-eight days and were therefore included in the death toll. It wasn't a large number.

Epidemiologists have a formal method for doing this kind of calculation: the attributable fraction. If you look at the mathematical formula for calculating the attributable fraction, you will see that it has to come out at less than the crude mortality rate. In other words, simply counting deaths exaggerates the problem.

On the other hand, there will be patients whose deaths are suspected to be due to novel coronavirus but they were never tested and so do not appear in the count based on cases who tested positive. For Scotland, if we include deaths of suspected but not confirmed cases – defined as having Covid-19 novel coronavirus mentioned on the death certificate – that increases the first wave count by 80%. For the second wave the figure is only 20% because testing was more widely available. Without knowing that you can't fairly compare the size of the two waves.

For all these reasons, many epidemiologists regard a metric called excess deaths as the gold standard. Excess deaths are calculated by comparing deaths in a given week of the year against an average of deaths for the same week in previous years. Unfortunately, excess deaths isn't a perfect measure either. One problem is that 2020 was an unusual year precisely because of novel coronavirus, unusual in ways that could well affect the underlying mortality rate, making comparisons with previous years unreliable. Perhaps more importantly, the excess deaths metric does not distinguish between deaths due to novel coronavirus and deaths attributable to the indirect harms of lockdown either. It weighs both sides of the scales at the same time. Counting deaths really isn't straightforward.

Hospitalisation is a particularly important outcome because it impacts on the NHS as well as on the patient. Hospital admissions are easy to count but that is too simplistic: not all admissions impose the same burden on our health care services. A more useful metric is hospital bed days, which can be weighted by the level of care provided, with stays in intensive care at the top end of the range. Even easier is to record the number of people in hospital with a positive novel coronavirus test result. This changes quite slowly – some people are in hospital for weeks – but it's indicative of the burden on the NHS at any one time. Mind you, all these simple measures will overestimate the problem to some extent; the attributable fraction problem that I described earlier applies no less to hospitalisations than it does to deaths.

The impact of illness in the community is hardest of all to quantify. Most people have mild infections and do not need to go to hospital, though they may be absent from work or unable to provide care to others. Acute illnesses lasting a week or two at most are not considered a significant health burden. That can change if there are serious long-term consequences – which the medics call sequelae – of an infection.

A number of sequelae of novel coronavirus infection have been bundled together as 'long covid'. An Office for National Statistics study in early 2021 estimated that as many as a one in seven of those infected suffered symptoms for at least twelve weeks. Long covid cases were given great weight by those trying to justify suppressing the virus at any cost, but their economic impact will be relatively modest. It could turn out to be less than

ME – a comparable condition of uncertain cause – that affects roughly a quarter of a million people in the UK. ME is cruelly debilitating, as I know full well from struggling with the condition for several years in my thirties. If long covid is anything like as bad then sufferers have my heartfelt sympathy. I am glad that support services are in place for long covid patients and that research is under way to find ways of preventing or treating the condition. I like to think that one day we will do the same for ME patients too. Let's be clear though, if long covid were the worst consequence of novel coronavirus infection then the idea of shutting down society to prevent cases would never have crossed our minds.

Some people have suggested that there may be other long-term consequences of infection that we don't yet know about. It will be impossible to rule that out entirely for many years yet, but nothing of this kind has been observed for other human coronaviruses, including SARS. This well-intentioned scare-mongering raises a broader issue: we can be lured into over-reacting to a new virus because we can't know what damage it might do. 'Better safe than sorry' is a reasonable starting point, but it shouldn't lead to our adopting costly or damaging interventions without some supporting evidence to justify our fears.

All in all, there are valid arguments both that the total health burden of novel coronavirus has been underestimated and that it has been overestimated. Resolving this may take some time, but the problem isn't much different than for a comparable disease such as influenza. For the rest of this book I shall use the UK's standard definition of novel coronavirus-related death – a death within twenty-eight days of a positive coronavirus test – and the equivalent for hospitalisations as measures of direct harms. They're not perfect, but they're useful. By July, when the UK's first wave had largely passed, the death toll stood at over forty thousand and more than one hundred and twenty thousand patients had spent time in hospital.

Harms caused by lockdown

The direct harms caused by novel coronavirus were substantial, and would have been worse if we hadn't intervened to slow the spread of infection, even if – as we saw in Chapter 3 – it's difficult to say precisely how much worse. However, it was equally clear that lockdown was going to be harmful too, and those harms must also be quantified if we are to make rational decisions about how to respond. Lockdown did indeed cause multiple harms. I shall review these under the five headings of health care provision, mental health, education, the economy and societal well-being.

One immediate harm caused by lockdown was a fall-off in health care provision throughout the UK. My colleague Aziz Sheikh was worried about health care provision from the outset and quickly set up a large project called EAVE to monitor it in Scotland. His data showed that visits to clinics, Accident & Emergency attendance, admissions of patients to hospital with conditions such as stroke or heart attack, routine operations and screening services all collapsed in late March and April 2020 after lockdown was implemented. They recovered slowly over subsequent weeks but still hadn't returned to pre-pandemic values six months later. There was a backlash – several politicians declared that never again should the NHS become a 'coronavirus-only' health service – but the damage was done.

It will take time to make a full assessment of the extent of that damage, but we can make a start. Some preliminary estimates of the number of premature deaths due to the fall-off in cancer screening alone exceeded ten thousand. In October 2020, the Office for National Statistics reported over twenty-five thousand more deaths at home than usual from conditions such as heart disease, cancer and dementia. Those people should have been in hospital, where some of them might have been saved.

It is hardly surprising that people stayed away from hospitals even when they desperately needed them; we were all being told that it was our duty to help prevent the NHS from being overwhelmed. So it was disconcerting to discover some months later that bed occupancy during the first wave (65% overall between April and June) was well below the long-term average (almost 90%). It was the small number of hospitals at close to 100% capacity that we saw nightly on the television news.

The impact of lockdown on mental health was a major concern from the outset, and there were worries too about domestic abuse and child safety. Several surveys monitored mental health during this period. Loneliness did increase, especially among those asked to shield, which led to changes to the shielding policy later in the pandemic. There were substantial increases in anxiety across the whole population, especially among younger adults, with women suffering more than men and those from deprived backgrounds most affected.

During the first lockdown, there weren't any marked increases in levels of clinical depression, so the short-term mental health harms of the first lockdown may not have been as bad as many had feared. That was to change in early 2021 when an Office for National Statistics study of longer-term effects reported substantial increases in depressive symptoms, more than doubling in young adults.

One of the most profound consequences of the first UK lockdown was that around ten million children missed an entire term of school and many missed important examinations. The imperative of getting children back to school was stressed by the Royal Society of London – the UK's leading science academy – and by the Royal College of Paediatrics and Child Health. A particular concern was that the children most affected were those already disadvantaged, exacerbating pre-existing inequalities.

It is hard to weigh long-term educational harms against the immediate public health problem that lockdown was intended to solve. What's unarguable is that we didn't react the same way during the swine flu pandemic of 2009–10. Schools were hotspots for swine flu and around seventy children in the UK died from infection, many more than died from novel coronavirus infection in 2020. Yet schools stayed open. It seems that our collective assessment of the balance of harms changed dramatically over the intervening ten years.

The economic harms caused by lockdown were more immediately visible and more easily quantified, whether in terms of gross domestic product (GDP), government borrowing, unemployment, bankruptcies or failed businesses and lost livelihoods. The UK economy contracted by an unprecedented 20% during the second quarter of 2020. Government borrowing rose by around three hundred billion pounds over the financial year. There were no published predictions of the likely economic impact before we entered lockdown and this gave the impression that we were prepared to try to suppress the spread of novel coronavirus at – quite literally – any cost at all.

Damage to the economy has long-term health implications – both because it puts the NHS budget at risk and because there is a known link between economic deprivation and poor health. This was almost completely ignored.

The fifth harm was that we were being asked to give up our social lives: family visits, weddings, funerals, performing arts, playing and watching sports, many hobbies and pastimes – everything that, for most of us, makes life worth living. Societal well-being is extremely hard to quantify (though there are metrics such as the happiness index), but it is accepted as a vital part of the human condition. The World Health Organization's constitution defines health as 'a state of complete physical, mental and social well-being and not merely the absence of disease or infirmity'. Yet in the first half of 2020 we effectively redefined health, more narrowly than ever before, as 'the absence of disease or infirmity caused by novel coronavirus'. Nothing else mattered.

Finding the right balance

I likened the problem of balancing harms to navigating through a six-dimensional maze. The dimensions were the harm to health due to novel coronavirus plus those due to the indirect impact of the epidemic on health care provision, mental health, education, the economy and societal well-being. Working out how best to navigate a way through the maze that minimises the net harm was always going to be extremely difficult. Mathematics can help, but only up to a point. How do you keep score when the metrics you are using are as wide-ranging as deaths, GDP and human happiness? In reality, finding the right balance is mostly subjective; it's a judgement call for the whole of society, not just scientists. My worry is that there were several factors in play that skewed those judgements and pushed us towards lockdown as the solution.

One of these was to caricature the debate as 'lives versus the economy'. There was only ever going to be one winner of an argument couched in those terms given the combination of an anxious public, politicians eager to be seen to be doing the right thing, and a government advisory system packed with clinicians and public health specialists. However, it was never the case that the downside to lockdown was purely economic, important though that is. This was also a question of lives versus lives.

There is an important rider to the lives versus lives issue. Most of the lives lost because of lockdown – such as premature deaths from cancer due to missed screening tests – will be lost in the future. This problem is well known to health economists and is normally dealt with by a technique known as discounting. Put simply, discounting allocates a value to deaths prevented in the future, while allocating a higher value to deaths prevented in the present. There are good reasons for doing the calculation this way, not least that the future may turn out differently from what we expect, but the explicit intent is to try to balance immediate harms against longer-term harms.

This standard health economics tool was jettisoned when we went into lockdown in 2020. There was no discounting at all of predicted novel coronavirus-related deaths up to two years in the future, even though the course of the epidemic was very uncertain indeed. At the same time, we discounted entirely – that is, we ignored – future deaths as a result of lockdown, as if their lives were of no value. This could only skew the argument in favour of lockdown, and it shouldn't have happened.

For good measure, the 'lives versus the economy' debate was twisted even further in favour of lockdown. The argument goes like this: novel coronavirus is bad for the economy, so suppressing the virus is good for the economy. This is such a naïve proposition that I wouldn't bother with it at all but for the fact it was advanced by some prominent economists. Yes, of course, suppressing the virus would be good for the economy, if it could be done without prohibitive cost. If we don't take into account the costs of an intervention we can end up with a cure that's worse than the disease.

Here's an illustration. Greenfly are bad for your rose bush, so suppressing the greenfly would be good for your rose bush. One way to kill off the greenfly is to use a flamethrower. Hopefully, over the years, the charred rose bush will recover. So you could argue that the strategy works – at least until the greenfly come back. That doesn't make it a good idea.

Another factor that skewed the debate in favour of lock-down was the commonly heard assumption that we would have a vaccine within a few months. This hope was bolstered by a series of encouraging bulletins from the many teams developing vaccines and by UK government announcements that it was pre-purchasing stocks. The problem is that the expectation of how long we will have to wait for a vaccine has a big influence on whether we think lockdown is a proportionate response.

My team did some work on this problem in March 2020, but you don't really need mathematics to work out the answer. Lockdown is not sustainable and the longer the lockdown the worse the damage. We could put up with lockdown for a short time and six weeks of full restrictions was talked about as the limit. No-one thought a vaccine was coming in six weeks, a year was about the best we could expect.

I think we would have taken a more sustainable approach to tackling novel coronavirus had it been widely understood and accepted that the epidemic would not be over in six weeks and that lockdown was a temporary fix not a permanent solution.

The alarming predictions from the mathematical models – often accurate but sometimes not – didn't help. The underlying problem was that the models were set up to explore the impact of a limited range of interventions on the dynamics of novel coronavirus cases, hospitalisations and deaths. That's all. SPI-M wasn't asked to balance the burden of novel coronavirus against the costs of those interventions or the harms they caused, and the models weren't designed to answer such questions anyway. It was implicitly assumed that lockdown came at no cost at all. As we saw earlier, if that were true then the best way to prevent deaths due to novel coronavirus was to go into lockdown and stay there.

All of this was entirely foreseeable. I had raised it with Catherine Calderwood, CMO Scotland, at a meeting in early

March 2020. The point I tried to get across at that meeting, and many others after, was that we needed to look beyond the public health burden caused by the virus. We failed to do so and ended up with a lop-sided body of evidence.

On the one hand we had a set of sophisticated predictions of how many cases, hospitalisations and deaths due to novel coronavirus we might expect under different scenarios. All of this was in the public domain.

On the other hand we had nothing to say about how much harm lockdown would do to the economy or anything else, nor about how many would die as a consequence of lockdown.

We epidemiologists were repeatedly told that this was someone else's job. Whose? Nothing was ever made public. Where was the formal impact assessment – a standard government planning tool – of the effect of lockdown? As I said to the House of Commons Science and Technology Select Committee in June: I think the novel coronavirus response is being driven too much by the epidemiology, and I'm an epidemiologist.

The verdict on lockdown

A great deal was said and written against the lockdown policy. In a letter to *The Times* in October, Matt Ridley and other peers from the House of Lords penned the memorable line: *If lockdown were a treatment undergoing a clinical trial, the trial would be halted because of the side effects.* Even that is understating the case given that we won't have a full accounting of the indirect harms of lockdown for many years to come.

Will the cure turn out to be worse than the disease? It is still early days for doing the reckoning but the omens are not good. As early as April 2020, the Office for National Statistics produced an interim report that was sent to SAGE. The report concentrated on indirect health impacts via damage done to health care provision and also to the economy, given that economic hardship and deprivation are well known to have adverse effects on health.

The report used a metric called quality adjusted life years (QALYs) to quantify both direct and indirect impacts of the epidemic. The QALY is the health economist's answer to the problem of combining death and illness in a single measure. As the measure is weighted by life expectancy, the impact on young people has a higher weighting than the impact on the elderly. The authors concluded that the QALYs lost to indirect effects were likely to be considerably more than the QALYs lost to the disease itself; the best estimate was three times more.

Even if future studies confirm this estimate – or, as I suspect they will, find an even greater imbalance – I must stress that this is not an argument for doing nothing in March 2020. The path through the maze was narrow and difficult to navigate; we needed to balance the indirect harms due to imposing measures such as lockdown against the direct harms of a worse novel coronavirus epidemic should we not do so. It is an argument, however, that we got the balance wrong, the March 23rd lockdown was too harsh and, if it had to be implemented at all, should have been at least partially relaxed much sooner than it was.

I recall one advisory group meeting where those of us favouring the relaxation of lockdown measures were characterised as being more willing to take risks. The comment was well intentioned but I strongly disagreed with it. I think we were all similarly concerned about the potential impact of novel coronavirus. The bigger difference was that some were more concerned than others about the risks of harm caused by lockdown.

Throughout 2020, many scientific advisors were all too willing to dismiss those indirect harms – several commented openly that their job was to minimise the harm done by novel coronavirus and nothing more. Surely our work should have been about finding a balance and, sadly, that meant accepting the reality that there would be harms one way or another. We were never going to get the balance right if our starting point was that 'no death from coronavirus is acceptable'.

CHAPTER 6

DANGEROUS LIAISONS

One of the first things we want to know about any infection is how you catch it. Novel coronavirus is a respiratory virus, like influenza virus and like the coronaviruses that cause common colds. All these viruses replicate in the upper respiratory tract and virus particles are expelled in droplets or aerosols as we exhale, cough or vocalise. We catch novel coronavirus by inhaling these airborne virus particles.

The airborne route is believed to be by far the most important way that novel coronavirus spreads but there are others. The virus can persist on surfaces – hence the advice for frequent cleaning early on in the pandemic – but it is now thought that this is not a major route of transmission. It may also be present in faeces (many respiratory viruses are also found in the gut), though that doesn't seem to be a common transmission route either.

Defining a contact

There are echoes of the novel coronavirus's evolutionary history in how and where it transmits most easily between humans. Novel coronavirus is thought to be descended from a bat coronavirus and its ancestors will have spread among huge colonies of bats roosting in close proximity in caves or mine shafts. Knowing this gives us some clues.

First, the ideal environment for the transmission of novel coronavirus is a poorly ventilated, enclosed space where it is still and dark, away from direct sunlight – 'cave-like' would be a fair description, but indoors works just fine.

Second, the virus spreads most efficiently wherever people are gathered together in close proximity for a prolonged period of time – much like bats roosting.

Third, the virus transmits especially well between people who are talking, shouting or singing – bats are noisy too. In normal circumstances, most people spend a lot of time indoors, close to others, and vocalising. This explains why – having made the jump from bats – novel coronavirus was able to thrive in human populations; it was predisposed to do so.

Understanding the biology of novel coronavirus helps us understand the concept of a 'contact'. A contact is any inter-action that provides an opportunity for the virus to spread from one person to another. The World Health Organization defined a contact in terms of distance apart and the time spent within that distance.

We don't have to be in direct contact with someone to catch novel coronavirus, but the further apart the better. There was early evidence that the risk of transmission was reduced by around 90% by being two metres apart, and 80% by just one. We don't have to be within that distance for long, but the fact that novel coronavirus spreads best where the virus is able to build up in still air implies that longer contact makes trans-mission more likely. The World Health Organization advised that a contact has to last at least fifteen minutes for there to be significant risk.

So, for practical purposes, a contact between people involves being no more than one or two metres apart for at least fifteen minutes. This is the basis for guidance to reduce the risk of transmission by physical distancing. The World

Health Organization initially recommended that people keep two metres apart wherever possible, later reducing this to one metre. The UK government settled on 'one metre plus'. The time dimension tended to get forgotten when guidance was issued, but it matters too.

Of course, not all encounters that satisfy the definition of a contact are equally risky. There will be a big difference in risk between spending fifteen minutes in a well-ventilated room wearing face coverings and sleeping in the same bed for an entire night – though both meet the criteria for a contact. Nor is anyone claiming that there is zero risk at five metres distance or from a fleeting encounter, just that the risk is too low to worry about. Not everyone accepted this, however.

The BBC News was fond of showing an animation illustrating how a single cough could infect everyone in a small restaurant. I complained to David Shukman – BBC Science Editor – about this. He replied politely enough but persisted with the animation. These are impossible arguments to win. I couldn't categorically state that such a thing could never happen, only that it had to be exceptionally rare. If novel coronavirus were so infectious that a single cough could routinely infect a roomful of people – or a bus-full or shop-full – then it would be unstoppable.

Novel coronavirus spreads particularly well within households. The bigger the household, the greater the risk. Visitors – especially overnight visitors – are just as problematic. This is one of the features of novel coronavirus that makes it exceptionally difficult to control. Workplaces and public spaces can be made safer, but there is little that governments can do to manage spread within our homes; that's up to us. What government can do is put an outright ban on social contact between households, which they did.

By far the safest place for people to get together is outdoors. As far back as March 2020 there was evidence from China that outdoor transmission of novel transmission was extremely rare. One study traced over seven thousand transmission events; just two occurred outdoors. For whatever reason, this knowledge did not influence UK government policy when it came to lockdown; we were told to stay at home and not meet others, even outside. There was no need for this, as long as physical distancing is observed the risk of transmission outdoors is negligible. Yet not only were sports such as golf and tennis banned – despite the low risk – but at one stage the police were even harassing solo hill walkers. That has to be one of the safest activities imaginable.

As the summer of 2020 wore on there were repeated outcries about crowds on beaches. People were castigated for their 'irresponsible' behaviour by politicians, police officers and public health experts. We were told to expect surges in cases. These never materialised, which was no surprise to anyone familiar with the epidemiological data. To my knowledge, no novel coronavirus outbreak has been linked to a beach anywhere in the world, ever.

There are a couple of caveats though. Intimate contact still poses a risk outdoors. Outdoor mass gatherings are problematic too. Large numbers of concert goers or sports fans sit or stand close to one another for a prolonged period and are often noisy (as we have seen, novel coronavirus transmission is linked to vocalisations). Even outdoors this combination may pose a risk, although as mass gatherings were one of the first activities to be banned there is little hard evidence. We should also be worried about so-called pinch points at such events – toilet facilities, indoor refreshments, and travel to and from the venue by car or public transport. All of these provide opportunities for the virus to spread.

Non-pharmaceutical interventions

There are three ways of preventing the spread of novel coronavirus by managing person-to-person contact. These are collectively referred to as non-pharmaceutical interventions because they do not involve any kind of medication.

First, there is a suite of measures that are intended to help reduce the risk of transmission associated with day-to-day activities. I call these Covid-safe measures and I'll describe them in more detail shortly. The point for now is that these interventions are about making contacts safe rather than not making contacts at all.

Second, there is self-isolation of cases and their contacts, an intervention that only affects people who have or are likely to have a novel coronavirus infection, thus minimising disruption to everyone else.

In contrast, the third approach – social distancing – is about reducing the number of everybody's contacts. The ultimate social distancing measure is lockdown.

Covid-safe measures include basic respiratory hygiene and hand-washing or sanitising. These measures were promoted by the UK government from the outset. Physical distancing, including the use of screens, was also quickly implemented where it was possible to do so. (Confusingly, this was often referred to as 'social distancing' – a term better reserved for measures such as banning household mixing, which is how I use it in this book.) Good ventilation improves Covid-safety, but fresh air is even better. Screening people for a high temperature was not recommended by SAGE and was not widely used in the UK, though it was adopted by some other countries.

One peculiarly controversial Covid-safe measure is wearing face coverings. It took some time for the UK to embrace this idea, which was odd because masks were always required for health care workers (who were, of course, at increased risk of exposure to the virus). There was speculation – but no evidence that I know of – that face coverings might engender a false sense of security leading people to behave less responsibly in other ways. This was akin to the argument in the 1980s that people wearing seat belts might drive more dangerously.

By July 2020 the advice changed and we were told to wear face coverings in enclosed public spaces. The UK was a bit slower off the mark than the World Health Organization, but even they didn't advocate general use of face coverings until early June.

Covid-safe measures do two things at the same time: they help protect you from any of your contacts should they be infectious, and they help protect your contacts from you should you be infectious. This makes it more difficult for the virus to spread and so reduces the R number. If we could reduce the average risk of transmission per contact by two-thirds (enough to reduce the R number to one from its natural maximum of three) using Covid-safe measures then we wouldn't have to do anything else to keep the virus under control.

To put this another way: if you could make contacts safe enough then there would be no need to cut the number of contacts you make, no need for social distancing and no need for lockdown. Unfortunately, it's not as easy as that. Relying on hygiene, physical distancing and face coverings alone is not enough to reduce the R number for novel coronavirus below one. Fortunately, we don't need to turn to social distancing right away, there are other ways to stop the virus spreading. First, though, we have to find it.

Outbreaks and super-spreaders

'Outbreak' is a familiar word (there was even a 1995 Warner Bros film of that name) but it is a technical term too. Public Health England defines an outbreak as two or more epidemiologically linked cases. That's a strict definition – most people don't think of two cases as an outbreak – but it is a standard and widely used one. 'Epidemiologically linked' means that public health workers can identify a time and a place where infection could have been transmitted – sharing a house would be an easy example. Two cases are not many though and here I will focus on larger outbreaks.

In 2020 epidemiologists at the London School of Hygiene and Tropical Medicine assembled a database of novel coronavirus outbreaks reported from around the world. By July they had amassed a list of well over two hundred, the majority with fewer than twenty cases but a few with over a thousand.

One striking feature of the London School's list is that few outbreaks occur outdoors. The ones that do are mostly associated with places of work – large farms or building sites – which raises the possibility of other links between the cases, such as shared eating places or sleeping accommodation. Consistent with what we know about the conditions that favour transmission, the great majority of large novel coronavirus outbreaks occur indoors in settings such as churches, hotels, factories, ships and care homes.

Digging a little deeper, we find different kinds of outbreak in the list. Some are linked to groups of people that frequent the same location, such as a workers' dormitory, university residence, school or workplace. Others are linked to one-off events, such as a wedding, bus trip or conference. This tells us that an outbreak results from a combination of setting, behaviour and perhaps characteristics of the individuals involved. Which brings us to super-spreaders.

The notion of super-spreading isn't new. In 1997 I wrote a paper showing that – for a variety of infectious diseases – some people contribute much more to transmission than others. A useful rule of thumb is that 80% of transmission is attributable to 20% of the population. A paper published in the journal *Science* in November 2020 reported that this rule worked well for novel coronavirus too. By analysing contact tracing data the researchers found that 20% of index cases were responsible for even more than 80% of secondary cases. We would like to know what is special about that small minority of cases. There are several possibilities.

First, some people will be more infectious because they shed larger quantities of virus, perhaps over an extended period of time. This is reminiscent of the sad story of Typhoid Mary who – back in the early twentieth century in the days before antibiotics – could not be cured of a persistent infection with typhoid-causing bacteria and was forced to spend much of her life in isolation to protect others. Persistent infections with novel coronavirus can occur – especially in people who are immunocompromised – but they are not common and, thankfully, we surely wouldn't react in the same way. But some variation in infectiousness is bound to occur, as it does for typhoid, influenza and many other infections.

Second, some people may have more contacts to whom they could pass on infection. A high number of contacts might, for example, be linked to working in public transport, hospitality, education or retail. For this reason, it might be a good strategy to target interventions at people in high-risk occupations or with high-risk behaviours.

The final possibility is that there is nothing special about the 20% other than the case happens to be – from the virus's point of view – in the right place at the right time. This brings us back to those one-off events that are associated with novel coronavirus outbreaks; I mentioned weddings, bus trips and conferences but it could be any occasion where large numbers of people spend time together in a confined, poorly ventilated space. Anyone present who happens to be highly infectious at the time – say, in the twenty-four hours before the onset of symptoms – could become a super-spreader.

In practice, where super-spreading has occurred – in a French ski chalet, an Israeli school or even (as has been claimed) the inauguration ceremony for a US Supreme Court judge – it is likely that a combination of all these factors played a role. For this reason, I have always been reluctant to apply the term 'super-spreader' to an individual. Super-spreading isn't just about an individual, it's also about how the individual fits into the wider epidemiological picture, who they have contact with and in what circumstances. I prefer the term super-spreader event.

The role of contact tracing

Ultimately, whatever factors are responsible for an outbreak the public health imperative is to contain it. That involves a practice called 'outbreak investigation', a mainstay of the public health response to infectious diseases that can be dated back to John Snow's landmark study of a cholera outbreak in London in 1854. Outbreak investigation was renamed 'cluster analysis' during the novel coronavirus pandemic. I couldn't see any reason for this, unless someone thought a nineteenth-century public health tool sounded too old-fashioned for a twenty-first-century pandemic.

Outbreak investigation combines identifying the source of the outbreak – a procedure known as backward tracing – with trying to identify those who may have been exposed during the outbreak – forward tracing. An important feature of tracing is that it is much easier to do when numbers of cases are low. Once there is widespread community transmission every outbreak is quickly absorbed into the rest of the epidemic and outbreak investigation loses a lot of its value.

Contact tracing is another routine public health activity. It is equivalent to forward tracing in an outbreak investigation but applied to a single index case. The aim is to identify every person who might have been infected by that case. In the UK, contacts of novel coronavirus cases were asked to self-isolate for fourteen days (reduced to ten days towards the end of 2020) and to report if they developed symptoms, at which point their contacts will be traced too. Members of the same household are almost always designated contacts. That makes sense: transmission within households is one of the most important ways that novel coronavirus and many other kinds of infection spread.

Contact tracing of this kind has to be done manually and has the disadvantage that it will only identify contacts or locations beyond the household that the index case is able to recall and is willing to divulge. For obvious reasons, willingness to divulge can be a huge problem, notoriously so in the context of sexually transmitted diseases. There is a short-cut available if the index case has been to an establishment keeping records (as many were required to do during the pandemic) that provide a ready-made source of potential contacts.

Speed is essential if contact tracing is to be as effective as possible. For tracing purposes, the infectious period is taken to be from forty-eight hours before symptoms appeared to seven days after. In practice, most transmissions are thought to occur in the first half of that period, which is why there is general agreement that contact tracing needs to happen fast. There is a lot to be gained if contacts are traced within a day but much less if it takes a week. This needs to be taken into account when tracing teams are overstretched; it will often be best to prioritise the most recent cases rather than trying to clear a backlog.

Mobile phone apps offer a fully automated – and fast – way of identifying contacts. In the first half of 2020 apps were promoted by some as an essential tool for tackling novel coronavirus. There was a lot of criticism of the UK government's inability to deliver one that worked. Enthusiasm had waned by the time the apps were finally rolled out in September and take-up was modest: about sixteen million people across the UK, just over one-quarter of the population. It would be bad enough if that low take-up made the app only one-quarter as effective, but the impact is far worse. A rough estimate of an app's effectiveness is the square of that fraction who use it. That's because both the case and the contact have to have the app for it to work. On that measure the apps in the UK were less than 10% as effective as they could have been. Even so, one study has estimated that self-isolations resulting from the use of apps saved several thousand lives in the second half of 2020.

Contact tracing apps bring us back to the definition of a contact. We saw earlier that a contact has to be within two metres for at least fifteen minutes, but the context matters: was the contact indoors or outdoors, was the room well-ventilated, were face coverings worn, was there physical intimacy? To an app, all contacts are equal but in terms of likelihood of transmission they are not equal at all, some kinds of 'contact' are thousands of times riskier than others.

How self-isolation works

Finding cases and contacts are important, but it's what happens next that reduces the spread of the virus. Self-isolation of cases and their contacts are at the heart of our response to a novel coronavirus epidemic. If self-isolation is working well then we have gone a long way towards winning the battle. It can have a major impact on transmission rates and is precisely targeted where the risk lies, so keeping disruption to wider society to a minimum.

We saw in Chapter 3 that the maximum R number for novel coronavirus in the UK in March 2020 was about three. That's when people who are infected behave normally: they may stay home if they feel unwell, or they may have no choice but to go out, or they may have only minor symptoms that don't alter their behaviour at all. If everyone with symptoms stayed home that would limit their contacts to members of their household, and they could be made safer by taking all precautions possible. By how much might that reduce R?

This answer depends on what fraction of transmissions occur before symptoms have appeared. That fraction is so important that it has its own name, sigma. The smaller the value of sigma the more likely any intervention targeted at symptomatic cases is to succeed. SARS had a low sigma value of about 0.10 (or 10%) and I'll use this to illustrate the argument. Sigma equals 0.10 means that if everyone with symptoms immediately and completely isolated themselves then the R number would be reduced by 90%, easily enough to bring it below one and control the epidemic. That's the key to how SARS was eradicated in 2003.

We were not so lucky in 2020. Sigma for novel coronavirus is considerably higher, close to 0.5, meaning that almost 50% of transmissions occur before symptoms appear. That implies that self-isolation triggered by the appearance of symptoms can never be enough to reduce R below one (from its maximum value of three). It gets worse. We also need to pay attention to the small print in the paragraph above: the words 'everyone', 'symptoms', 'immediately' and 'completely'. Let's look at these one by one.

Everyone: for one reason or another not everyone will choose to or be able to self-isolate.

Symptoms: not everyone will recognise that they have symptoms and some infected people may not even develop symptoms at all.

Immediately: some people may wait 'to be sure', or wait until they have a positive test result.

Completely: anyone living in a multi-person household or needing care will find it difficult to isolate themselves fully from other household member, as will anyone requiring care. The upshot is that, in practice, only a modest reduction in R can be achieved by self-isolation of symptomatic novel coronavirus cases.

Fortunately, we can improve matters by contact tracing and having contacts of cases – including members of the same household – self-isolate too. In principle, self-isolation of contacts is more effective than self-isolation of cases. That's because if it begins promptly it can reduce pre-symptomatic transmission too. It will have to be prompt though. If the contact is not traced until four or five days after they were exposed – and remember there will be a delay while the index case is reported and confirmed – then they may already be infectious and infecting others.

There is still the question of how many of the contacts are traced in the first place. By September 2020 the UK government was publishing this statistic every week and it was taken as a key indicator of how well the entire system was working. The target set was 80% of contacts reached; this was often met and sometimes exceeded. That isn't quite as good as it sounds because it includes household contacts, and you don't need an elaborate tracing system to identify household contacts. The requirement for members of the same household as a case to self-isolate had been introduced even before we went into the March 23rd lockdown.

Putting all this together in a mathematical model, a report from the Royal Society of London estimated the combined effectiveness of contact tracing and self-isolation of cases and contacts in terms of reducing the R number. They came up with a figure that was not enough to bring R below one, but would get us well over half-way there. That is why I said earlier that if contact tracing is working well – and people do self-isolate when they need to – then we can make good progress in reducing the R number.

There is one rider, however. The 80% target for contact tracing ignores a critical element of contact tracing – the fraction of cases that are recognised and reported in the first place. As we shall see later, case finding was far from perfect and this had a significant impact.

This evidence tells us that we need to pay close attention indeed to how well self-isolation is working in practice. If it isn't working well we have two options: try to improve matters, or bring in additional restrictions.

That is why I was extremely concerned when a study called CORSAIR was published in September 2020. The CORSAIR study reported that, although most people said that they fully intended to adhere to the self-isolation rules, few managed to do so: the headline figure was less than 20%. There is an important caveat that not adhering meant not adhering perfectly. Many breaches may have been minor, so this does not mean that self-isolation was only 20% effective. It will have been better – perhaps much better – than that. Even so, I feel this report should have received far more attention than it did. Instead, when cases started to rise later that same month, the talk was not of improving adherence to self-isolation, it was all about lockdown.

Behaviour matters

As we've seen, the transmission of novel coronavirus is directly related to our behaviour – the number and nature of close contacts we make with others. During the March 23rd lockdown the R number fell to around 0.7, less than a quarter of its natural maximum. As various surveys showed, a large part of this reduction was due to people changing their behaviour.

Comix surveys showed that we had reduced our average number of contacts from over ten to less than three per day. Mobility data collected anonymously from people's phones showed that we moved around much less, spending much more of our time in residential areas, not going to work or travelling further afield. High percentages of people reported wearing face coverings and adhering to social distancing. As you'd expect, we also spend more time outdoors in the warmer months; that helps because novel coronavirus does not transmit well outdoors.

Stephen Reicher – a behavioural scientist at the University of St Andrews – argued that changes in people's behaviour were motivated not only by a desire to keep themselves safe but also to protect others, that this was a community enterprise. I have to say that many of the behavioural scientists I spoke to in 2020 had a more positive view of human nature than I'd be comfortable assuming, but I hoped Stephen was right. The battle with novel coronavirus had to be a community enterprise. The most important elements of our response – case reporting and self-isolation of cases and contacts, which involve considerable self-sacrifice – are not for the benefit of the cases themselves, they are for the benefit of everyone they might otherwise come in contact with.

UK administrations did not, however, stake everything on our community spirit, they also introduced sanctions and penalties. Many of my behavioural science colleagues felt that there was too much emphasis on penalising those who broke the rules and not enough on providing help and support to those that needed it.

For me, the main problem with the ways the rules were implemented was the danger of everyone concerned losing sight of what was most important. Hill walkers and beach goers are not a threat to public health; anyone with symptoms of Covid-19 who does not get tested and self-isolate most certainly is. That message was diluted by failing to focus on the behaviours that posed the greatest risk.

The situation wasn't helped by a series of incidents where high-profile figures were caught flouting the rules they themselves were responsible for. First there was Catherine Calderwood, CMO Scotland. She was followed by Neil Ferguson, SAGE member. Then it was Dominic Cummings, senior advisor to the PM, and finally Margaret Ferrier, Scottish National Party MP. A team at University College London found that the Cummings incident coincided with a marked fall-off in the public's confidence in the UK government and affected willingness to adhere to the rules as well.

I should add that I was accused of rule-breaking myself. Somebody – I don't know who – decided to tell a tabloid journalist that I'd moved my family to the Western Isles of Scotland after the March 23rd lockdown was imposed. This was easily disproved: we'd made a long-planned trip a few days before and – like many thousands of others – got caught by the lockdown. Admittedly, this wasn't my most impressive piece of pandemic forecasting, though – as one sympathetic colleague later remarked – one of the hardest tasks we had in 2020 was predicting what the government was going to do and when.

I was directed to some very capable solicitors who stopped the false claim being printed, but the tabloid published a nasty little story anyway implying we were not welcome in a community I'd been part of for forty years. Any ill will was in short supply. Instead, we received a stream of kind and greatly appreciated messages of support, accompanied by gifts of eggs, vegetables and jams. Our sheep-farming neighbour – one of my oldest friends – loaned my teenage daughter three pet lambs to look after and our week-long stay ended up lasting four-and-a-half months, by which time the lambs had grown a lot bigger and seemed to think they had the run of the house as well as the garden.

A problem we all faced during 2020 was the sheer complexity and ever-changing nature of the rules we were supposed to live by. Rules can only have the desired impact if people know what they are, understand what they mean and can follow them. I lived in constant fear of being asked in a media interview what FACTS stood for and failing to remember. FACTS was Scotland's equivalent of 'hands, face, space' in England and stood for face coverings, avoid crowds, clean hands, two metre distance and self-isolate. Polls showed that only a tiny minority could recall the whole set. There were repeated calls for clearer public health messaging but we never got it.

How we behave is, of course, influenced by what government tells us to do, but that will not be the only factor. Our personal circumstances inevitably affect what we are able or willing to do and, perhaps most important of all, our behaviour will be driven by our perception of the risk to us and those around us. To assess those risks properly we need to understand not only how novel coronavirus spreads – which was the subject of this chapter – but also the consequences of getting infected – which is the subject of the next.

CHAPTER 7

A DISEASE OF OLD AGE

As early as February 2020, reports from China showed that the risks of severe illness and death from a novel coronavirus infection increased sharply with age. I spoke to the popular science magazine *New Scientist* about this in early March but it wasn't until several weeks later that the pattern become widely appreciated, thanks mainly to the work of Carl Heneghan at the University of Oxford and David Spiegelhalter at the University of Cambridge.

Age as a risk factor

There is a simple way to illustrate this pattern of age-related risk. The average age at death in the UK is seventy-eight years. According to the Office for National Statistics, the average age of deaths attributed to novel coronavirus infection up to October 2020 was eighty years. In other words, novel coronavirus-related deaths were more skewed to the elderly than were deaths from any other cause. I'd say that was a reasonable definition of a disease of old age. Some causes of death – dementia is one example – are even more skewed to the elderly but it is an unusual pattern for an infectious disease. There's only one comparable example: SARS, novel coronavirus's closest relative.

During the UK's first wave more than 80% of the forty thousand or more novel coronavirus-related deaths were in people over seventy years old, an age group making up just 15% of the population. Those over seventy years old had at least ten thousand times the risk of dying as those under fifteen years old. I will repeat that sentence because it has such huge implications: those over seventy years old had at least *ten thousand* times the risk of dying as those under fifteen years old.

There is also a strong age bias in hospitalisation rates, the metric that directly relates to NHS capacity. More than one hundred and twenty thousand patients were hospitalised with Covid-19 during the UK's first wave: their median age was about seventy years. This is older than the median age of patients hospitalised from all other reasons – about sixty years – but it does mean that almost half of Covid-19 patients in hospital were under seventy years old (another name for the median is the 50-percentile, meaning the exact middle of the distribution). Even among hospitalised patients, though, the risk of dying increases markedly with age.

Older people are more likely to die from novel coronavirus infection or be hospitalised but they are not more likely to be infected in the first place. This didn't become fully apparent until testing and surveillance improved over the summer but we now know that teenagers and younger adults were at least twice as likely to be infected than the over-seventies. That's no surprise: it reflects well-known differences in contact rates between the different age groups. The implication of lower infection rates in the over-seventies is that the infection fatality rate – the risk of dying given that you have been infected – is even more biased toward the elderly than a simple count of the number of deaths indicates. Covid-19 really is disproportionately a disease of old age.

More risk factors

Age is by far the most important determinant of risk from novel coronavirus infection but it isn't the only one. Thanks to a huge amount of painstaking analysis by epidemiologists around the world we now have a long list of attributes that are associated with severe illness or death. We call these attributes risk factors. Known risk factors include a range of chronic, underlying health conditions such as diabetes, chronic obstructive pulmonary disease, kidney disease, dementia, obesity and being on chemotherapy. These are referred to as co-morbidities. None of them – with the exception of Down's syndrome – has as big an impact on risk as age.

It is not as simple as that, however. Co-morbidities tend to accumulate with age. This is the reason why it's often so difficult to attribute a death to a single cause. Could this explain the age effect? The answer is a firm no. The co-morbidities do make a difference but, even when this is accounted for in statistical analysis, age remains by far the most important risk factor. This leaves open the question of what exactly it is about age that matters. One suggestion is immune-senescence, the gradual deterioration of the body's ability to combat any infection, but we don't yet know precisely how this might work.

Age and co-morbidities are not the only risk factors. Being male is a risk factor, as is being non-Caucasian or having certain occupations (with health care workers near the top of that list). Some of these will be risk factors for getting seriously ill if infected, some will be risk factors for getting infected in the first place, and some may be both.

To complicate matters further, some of these risk factors are – like co-morbidities – associated with age, some are associated with one or more co-morbidities, and some – such as gender and occupation – are associated with each other. Teasing out the contribution of so many different and inter-related risk factors is far from straightforward. This doesn't only matter to epidemiologists; it matters to all of us if we want to assess the risk to ourselves as individuals.

There are various ways to express individual risk as a single number. One approach is an index of 'frailty'; several versions are used in geriatric medicine. Another is to work out your 'Covid Age', which is your age with years added for known risk factors such as being male or having diabetes. Another is simply to count the number of known risk factors a person has. Those metrics are helpful but they don't tell you what you most want to know: what's the probability that you will go to hospital or die from a novel coronavirus infection?

There are several different algorithms available that calculate individual risk of death or severe illness from novel coronavirus based on a selection of the known risk factors. It is easy to find these online, although I don't recommend taking the results too seriously unless it is clear that the methodology is sound – as we've seen, risk calculations are not straightforward. I tried one of these algorithms during the summer of 2020. It assessed me as having a 1% chance of dying if I got infected with novel coronavirus. I'm in good health but being over sixty and being male increased my risk substantially.

Most of us are not that good at assessing risk. How concerned should you be that if you caught a novel coronavirus infection your risk of dying from it is 1%, or one in a hundred? That's a personal question and each of us will answer it differently. For me, one in a hundred is enough to take precautions against getting infected for my own sake as well as for the common good. What if it were one in a thousand or one in a million? Some may regard all those numbers as too small to worry about. Others may worry about any risk that isn't zero.

David Spiegelhalter did an analysis showing that if you

get infected with novel coronavirus your risk of dying closely tracks your risk of dying anyway over the next twelve months. You may not know your risk of dying but actuaries do this kind of calculation every day. You can be sure that your life insurance provider has an accurate idea of the odds you will live to pay your next premium. David's analysis suggests those odds are a good indicator how worried you should be about novel coronavirus.

Sally Davies – ex-CMO England – took this idea one step further by proposing that the UK suffered badly from Covid-19 because we have an unhealthier population than many other European countries. I agree with Sally that this could – at least partly – account for the UK's relatively high mortality rate during the pandemic. This is especially true for Scotland, which has the lowest life expectancy in western Europe.

In October, a team led by Julia Hippisly-Cox at the University of Oxford published a risk estimation algorithm called QCOVID. The team analysed data from over six million people and validated their algorithm by testing its predictions with data from over two million more. The results were remarkable. The 5% of people predicted to be at greatest risk accounted for a staggering three-quarters of all deaths attributed to Covid-19. The QCOVID study strongly supported what we already suspected; the risk of dying from novel coronavirus infection is hugely concentrated in a small subset of the population.

Thanks to QCOVID, we can identify that subset with much greater precision. We saw earlier that 80% of novel coronavirus-related deaths occurred in the oldest 15% of the population. According to QCOVID, 91% of deaths occurred in the 15% of the population at greatest risk. What has happened is that we have removed the healthy elderly from the top 15% and replaced them with younger individuals with co-morbidities and other risk factors.

We could continue further down the risk rankings: between a third and a half of the population have at least one co-morbidity that increases their risk of dying from novel coronavirus infection. Obesity is one of the most common. We would then have accounted for the vast majority of novel coronavirus-related deaths, but not quite all. There have been several hundred deaths from a novel coronavirus infection among healthy adults under sixty years old. We don't know why these apparently random deaths happen. One idea is that some people have a genetic variation that weakens the specific part of their immune system responsible for responding to virus infections. If that is confirmed then we can refine our list of risk factors accordingly.

In the meantime, novel coronavirus-related deaths in healthy young and middle-aged adults – however rare – received a disproportionate amount of media attention and so had a huge impact on public perception of the threat. They were often used to justify lockdown.

Another device that was used to justify lockdown was to exaggerate the expected reduction in life expectancy. There's a simple calculation called years of life lost (YLL) that weights each death by the life expectancy of a person of the same age. Several studies estimated this at ten years for every Covid-19 patient who died, adding up to a colossal loss of life years. This didn't sound right to me: many of those who died were care home residents, and the average life expectancy of a care home resident is about two years. On closer inspection, it turned out that these studies treated everyone of a given age equally, ignoring the fact that underlying frailty and infirmity are key drivers of mortality due to Covid-19, and so substantially exaggerating the loss of life years.

The blunt truth is that we would never have contemplated lockdown as a response to novel coronavirus if no-one was at greater risk than a healthy sixty-year-old. We would still have demanded a vigorous public health response: it's a serious disease and, if left unchecked, has the potential to cause illness in a huge number of people, putting immense strain on the NHS. Even so, it's the much higher hospitalisation and mortality rates in the elderly, frail and infirm that forces us to contemplate lockdown. So it is fair to ask if there were any other ways to reduce that burden of death and disease.

Protecting the vulnerable

There are several ways we can protect those most vulnerable to novel coronavirus. There are vulnerable categories such as hospital patients and care home residents that we can protect by making hospitals and care homes as Covid-safe as possible. Then there are millions of vulnerable people in the wider community who we can try to protect individually. Let's look at each of these in turn.

The prospect of novel coronavirus spreading in hospitals was a concern from the outset. Infection control in a hospital is a perpetual challenge; superbugs such as MRSA continue to take their toll of hospital patients around the world. We knew that the SARS virus, the closest relation to novel coronavirus, transmitted readily within hospitals. So does another dangerous coronavirus, MERS. There was evidence of nosocomial transmission – the technical term for spread within hospitals – of novel coronavirus in Wuhan, China, from early on in the pandemic.

For these reasons, measures were quickly implemented to protect patients and staff in UK hospitals. Even so, according to one report, as many as a quarter of Covid-19 cases in hospital during the UK's first wave were infected while in hospital. These patients were much more likely to be elderly, frail or infirm – and therefore at higher risk of dying – than those who acquired their infection in the community.

It wasn't only the patients who were affected; there were reports of hundreds of novel coronavirus-related deaths among hospital staff, both clinical and non-clinical. All staff were in danger of being infected by Covid-19 patients and some hospital procedures – such as putting a patient on a ventilator – put clinicians at particularly high risk.

The virus didn't only pass from patients to staff – we shouldn't underestimate the importance of the staff-to-staff route. One of my medical colleagues believes that more staff got infected when relaxing in the canteen than they did on the wards, where precautions are routine. However hospital staff were exposed, illness, infections and suspected infections led to significant personnel shortages, another way that the virus affected health care provision.

There were clearly problems with the UK-wide response to novel coronavirus in hospitals, not least the logistic failure early in the pandemic to procure adequate supplies of personal protective equipment (PPE). Yet no-one could say that the risk to both patients and staff in hospitals was not widely recognised from the outset, prioritised and acted upon (though it took a little while to realise that 'staff' included support staff – such as cleaners – as well as nurses and clinicians). The phrase 'Protect the NHS' had multiple meanings – it also meant keeping patients away – but it was on everyone's lips. Despite this heightened awareness, hospital-acquired infections made a significant contribution to the public health burden of novel coronavirus in the UK, so it's hard to argue that we did enough.

We certainly didn't do enough in care homes. By the end of the first wave more than a quarter of the forty thousand novel coronavirus-related deaths in the UK were among care home residents. In Scotland it was over 40%. These are striking numbers given that fewer than half a million people in the UK – less than 1% of the population – live in care homes. It is widely accepted that the UK failed to protect care home residents during the first wave of the novel coronavirus epidemic. The question is why that failure was allowed to happen.

The clues were there for all to see. We'd known since February that novel coronavirus was a far greater threat to the elderly. We also knew that it spread well in residential facilities: one high-profile example was a cruise ship, the Diamond Princess. The Diamond Princess ended up in quarantine in Japan for most of February after a large outbreak of novel coronavirus. There were over seven hundred cases and fourteen of the mainly elderly passengers died.

The word used to describe precautions taken to keep a virus – or any other cause of infection – out of a care home or any other facility is 'biosecurity'. Yet even the most basic biosecurity measures weren't taken. Patients were discharged from hospitals back to care homes without being tested for novel coronavirus, even though hospitals were known to be hotspots of transmission. PPE was not always reliably available. Staff moved from one care home to another in the usual way of the industry without waiting to find out if they had symptoms or tested positive.

Medics and public health practitioners are concerned about infection control in hospitals, but the word 'biosecurity' isn't part of their day-to-day vocabulary. It's a word most would associate with a biological laboratory facility such as Porton Down rather than a care home. Vets, on the other hand, talk about biosecurity all the time. The livestock industry depends on protecting the health of the whole herd or flock, not just individuals. As the care home crisis unfolded I received e-mails from veterinary colleagues shocked by what was going on and wondering if they could help in some way.

The Health Secretary, Matt Hancock, acknowledged the risk to care home residents in Parliament on March 3rd and SAGE did the same in a meeting on March 10th. This wasn't forceful enough. Action taken at that time could have made a big difference – the first cases in care homes in the UK weren't reported until March 12th. When I later asked one scientific advisor – who was a vocal critic of government in the early stages of the pandemic – why they had never once mentioned biosecurity for care homes the reply was that 'the modellers should have thought of that', an unconvincing answer from an expert in public health.

Mind you, I wasn't talking enough about care homes myself either. I was still trying to get the message across that the elderly, frail and infirm were at hugely increased risk – this was not widely understood or accepted at the time. I do now feel that I was too concerned with the bigger picture and could have placed more emphasis on care homes specifically.

Advisors not highlighting the threat was one issue. Another was that political attention during March 2020 was on protecting the NHS and ensuring it had the capacity to deal with the expected influx of Covid-19 patients. Little or no attention was paid to the needs of social care. The transfer of hospital patients back to care homes was reportedly motivated by the need to free up beds, any risk to the care homes being a secondary concern. In mid-March, government testing capacity – which was limited at the time – was diverted entirely to hospitals; testing in care homes wasn't prioritised until it was too late. It was right to be concerned about hospitals, but not to the exclusion of everything else.

Hospitals and care homes were hotspots for novel coronavirus but we shouldn't forget that the majority of the people who died during the UK's first wave acquired their infections in the community, many of them in their own homes. From March 2020, the UK's strategy for protecting vulnerable people in the community was termed 'shielding'. The premise was reasonable enough: if certain individuals were thought to be especially vulnerable to novel coronavirus then it made good sense to tell them to take extra precautions in the face of a rapidly growing epidemic.

The importance of shielding

Over two million vulnerable people in the community were told to shield themselves during the first wave in the UK. The entries on the list of conditions that triggered shielding advice were informed guesses at the time but included some (though not all) of the co-morbidities now confirmed to be important, including chronic obstructive pulmonary disease and being on chemotherapy. Everyone over seventy years old was advised to take extra care but was not told to shield.

The big problem with shielding was what vulnerable people were told to do: self-isolate and not leave their homes. Doing this was challenging for both practical and psychological reasons and shielding in this way for a prolonged period was neither sustainable nor desirable. Before long, the shielding policy had become irrevocably associated with the misery of extreme self-isolation and deeply unpopular.

At the time, it was not known how well shielding was working either. No good estimates of the impact of shielding were made in 2020. We did know that, even with shielding in place, novel coronavirus-related deaths were hugely concentrated in the most vulnerable groups that it was supposed to protect. However many lives shielding saved, by itself it was nowhere near enough to protect vulnerable people in the community. Throughout 2020, however, none of the UK administrations prioritised protecting the vulnerable in the community. Even when it was mentioned it always felt like an afterthought. Before suggesting what more could have been done, we might ask why this was allowed to happen in the first place. I can think of three reasons.

The first is to do with epidemiological jargon. During the pandemic epidemiologists spoke constantly of the need to reduce the contact rate, which is the number of people we meet each day in circumstances where the virus could pass from one to the other. The problem is that we didn't really mean contact rate, we meant transmission rate. From an epidemiological perspective halving the risk of transmission per contact is precisely equivalent to halving the number of contacts.

The focus on contact rate led to the idea of shielding people by telling them to self-isolate. Though it certainly helps if vulnerable people minimise the number of contacts they make – as we were all asked to do during the pandemic – we can also protect the vulnerable by making those contacts safer. This wasn't made clear, so policy-makers continued to think of shielding in terms of self-isolation, not Covid-safety.

The second reason was fatalism. I was told on numerous occasions that implementing Covid-safe measures to protect the vulnerable in the community was just too difficult. Yes, the scale is daunting – we are talking about several million people – but surely no more daunting than locking down the entire country, and we were all too willing to do that.

The combination of unhelpful jargon and fatalistic attitudes didn't help, but I don't think the argument that the vulnerable couldn't be protected – or, if they could, then only by means of extreme self-isolation – would have held sway for the whole of 2020 if there hadn't been a third factor in play. Protecting the vulnerable was seen by some as challenging what I came to call our obsession with suppression. Let me explain.

The main justification for lockdown is that it suppresses the virus, driving down levels of infection, and that this makes everyone safer, including care home residents and the vulnerable in the community. This is true, but it's not enough. We saw in Chapter 4 that at least half of deaths during the first wave were the outcome of infections acquired *after* lockdown was imposed on March 23rd. Many of these deaths were care home residents but the majority were in the wider community.

Protecting the vulnerable is fully compatible with suppressing the virus – you can do both. I'd thought that was perfectly obvious, but it turned out it wasn't. I attended a top-level Scottish government briefing where one scientific advisor proclaimed with great conviction that you couldn't, that the two were incompatible. This was ridiculous and I tried – as diplomatically as I could given who was in the virtual room – to correct the misimpression, but I fear the damage was done. From that meeting on, protecting the vulnerable came a distant second to suppressing the virus in Scotland. The argument for doing more is straightforward enough, on this issue there's no balance to get right: the better we protected the vulnerable, the fewer people would die.

My team made a two-fold contribution to demonstrating the key role that protecting the vulnerable could play in our response to novel coronavirus. We showed that protection was easier to achieve and need not involve self-isolation if everyone who comes in contact with a vulnerable person is protected too. We went on to show that more effective protection of the vulnerable and their contacts could reduce the need to suppress the virus in the general population.

To explain how this works I first need to explain segmenting. Segmenting means directing measures at a subset of the population rather than the whole population. The subset might be the two million told to shield during the first wave, or the three million at greatest risk as identified by QCOVID, or the five million or more who are over seventy-five years old. Put simply, it is a risk-based approach.

Measures directed at those who are at highest risk have several advantages: they protect the people who most need protecting and reduce the burden on the health system, while minimising the indirect harms suffered by the whole population. A risk-based approach was always going to be the most direct and efficient way to deliver the UK administrations' policy objectives of saving lives and protecting the NHS.

My team took segmentation a step further by introducing the idea of a segment of 'shielders'. Shielders are the community equivalent of care home workers. They are the closest contacts of vulnerable people: members of the same household, care-in-the-home workers, informal carers and other essential visitors.

Shielders make up a large segment of the population. Even just counting carers, across the UK there are over one and a half million social care workers and more than four million informal carers. A vulnerable person is at greatest risk of infection from their shielders. It may not be possible – it certainly isn't desirable – to cut off contacts with the shielders. For that reason, it is vital to make contacts with them as safe as possible. There are three ways of doing that.

First, contacts with the shielder can be made safer by taking every practicable Covid-safe precaution: hygiene, PPE, ventilation or meeting outdoors, and physical distancing. This isn't normal life, but it's surely better than self-isolation.

Second, there is testing. Screening for symptoms is important but it's not enough because of the preponderance of pre-symptomatic and asymptomatic transmission. Regular testing – at least twice weekly – can be used to establish that shielders are free of infection. Should a shielder have symptoms or test positive, however, there has to be a response. If the shielder is co-habiting with the vulnerable person then they may need alternative accommodation. If the shielder provides care then that role needs to be filled by someone else while they self-isolate. Government or local authorities could help provide both kinds of support in principle, but to do so would require funding and proper planning for delivery at scale.

A third way to make contacts safe is for the shielders to take care to minimise their own risk, thereby reducing the risk to the person they are shielding as well. This puts a lot of responsibility on the shielders but it does provide the vulnerable with a buffer against infection circulating in the wider community. Exactly the same argument applies to care home and health care workers too; minimising the risk of their becoming infected reduces the risk to those they are looking after. In medical parlance this strategy is called cocooning.

Cocooning requires that a shielder is prepared to forego or make safe their social or work contacts to reduce the risk that they become infected and unwittingly infect the person they are shielding. That may be difficult to sustain for a long period; for example, if a carer wants to visit their own children or has to go to their workplace. Those activities are made safer if the children or the workmates are themselves behaving responsibly and are cautious about having symptoms, about having been in contact with anyone potentially infected and about having been in a high-risk environment such as a university residence or a hospital. I call this the 'chain of trust'.

One person who had worked all of this out for herself – and inspired my thinking on the subject – was my sister. She is my mother's informal carer and her closest companion. As early as February 2020, my sister realised that she needed to protect herself so as to protect my mother. Though the term didn't exist at the time, she had identified herself as a shielder. That meant the rest of my sister's family needed to take extra care not to put my sister at risk so that she didn't put my mother at risk – the chain of trust. They steadfastly kept this up and successfully saw the pandemic through to the point where my mother received her vaccinations in early 2021.

My sister – and untold thousands of others in similar circumstances – did all this without any outside assistance. Throughout 2020, the UK administrations failed to grasp the importance of the shielder. The nearest they came was the concept of the 'support bubble' (though bubbles were not primarily about protecting the vulnerable). Shielders were not given any specific advice, help or support. For example, they were not provided with PPE, nor were they were given special access to testing. It wasn't until early 2021 when vaccination priority lists were being compiled that the importance of protecting shielders was finally recognised.

The combination of measures I've described to protect vulnerable people in their own homes are similar to the measures needed to protect care home residents. I've simply taken the concept of biosecurity for a care home and applied it to households. This approach was always an option – I do not accept the argument that there was nothing we could have done beyond shielding by self-isolation.

In modelling segmentation and shielding my team was ploughing a lone furrow. None of the other SPI-M models included contacts between vulnerable persons and their shielders so they were not set up to study the potential of this approach. Our results were encouraging though. We showed it was possible to decouple the burden of deaths and severe illness from the R number: the more effective the segmentation and shielding, the more restrictions can be relaxed for the population as a whole without increasing the public health burden.

A segmentation and shielding approach gives policy-makers the option of relaxing more measures than they would otherwise feel able to do without overwhelming the NHS. Alternatively, they may choose to keep the measures, suppressing the virus and maintaining the R number below one, but reduce the mortality rate at the same time. We thought these were good options to have.

We published our write-up of this work on a website called medRχiv in early May. medRχiv publishes what are known as pre-prints, scientific papers that have not yet been peer-reviewed and accepted for publication in a scientific journal. Publication in a journal takes time and we felt this work was too important to wait. On the other hand, because it was important, we also didn't want to release it until it had been peer-reviewed; that is, the paper had been read and approved by independent experts in the field. We solved our dilemma by taking advantage of a rapid review service for novel coronavirus papers provided by the Royal Society of London. We published the pre-print immediately after this process was completed.

The pre-print generated a huge amount of interest because it promised a quicker route out of lockdown, which was on everyone's mind in May 2020. The Science Media Centre in London arranged for us to give a press briefing, the story was carried by most major newspapers and I was asked to talk about it on Radio 4's *Today* programme. More importantly, our report attracted a lot of interest from policy-makers. I discussed it with the DHSC, Scottish Government, the Treasury and the Cabinet Office, and I'm told that a copy went to the Prime Minister's Office.

Despite all this attention, the idea did not take off. Part of the reason it didn't was that I had failed to pay enough attention to language. We used the word 'shielding' in the title and, as we saw earlier, shielding had become unpopular through being linked to the government's approach of extreme self-isolation, which was now being spoken of as unethical even by some government scientists.

This was one of a handful of occasions when I received significant amounts of hate mail during the pandemic, all from over-seventies angry that – as they saw it – I wanted to force them into indefinite self-isolation. That was not what we had in mind; our approach was to make contacts between vulnerable persons and shielders safe, not to stop them having contacts at all. Another criticism of our pre-print was that we were advocating a do-nothing approach in the general population. This one wasn't true either, we made it clear that some measures would still be necessary to reduce the transmission rate in the community, but once the charge was made it stuck.

Several people have suggested to me that the paper wasn't given a fair hearing because it challenged the orthodoxy that suppressing the virus was the only viable approach to managing the epidemic. Maybe that was the problem, or maybe we simply didn't make the case well enough.

Playing the ethics card

The debate about shielding became heated and passionate. As the rhetoric intensified some truly bizarre claims emerged. One senior medic wrote that *the idea of selective protection of the elderly and the vulnerable is unethical*. This statement was made in the context of a kind of herd immunity strategy and I know full well – because I asked – that the writer was entirely comfortable with our approach to segmentation and shielding. Still, it was an extraordinary statement to read: how can protecting people who need protecting be unethical?

Cultural perspectives matter too. In Africa, for example, shielding is seen as fully consistent with the principles of *ubuntu* or 'I exist because of we'. A team of Ethiopian academics wrote in the *British Medical Journal* in July 2020 that shielding was seen as a way for an entire community to help protect the vulnerable (an extended version of the chain of trust), and far more acceptable than lockdown. Back in the UK there are minorities – peanut allergy sufferers are a good example – who society tries to protect from risks that they face but the majority of us do not.

In an attempt to resolve the ethics issue I wrote to Mike Parker, director of an ethics institute at the University of Oxford and a member of SAGE. Mike was clear in stating that there was no way of responding to the novel coronavirus pandemic that didn't have different impacts on different segments of the population. He felt that shielding in some form would prove to be necessary. That resolved the issue for me, but it never went away.

Mike's point about different impacts on different segments raises the issue of inequity. The health burden caused by novel coronavirus falls most heavily on a small minority of the population. The burden of lockdown also falls more heavily on some people than others, such as those through their personal circumstances denied access to work, childcare or education. It is an awkward truth that these are not the same groups: the people who benefitted the most from suppressing the virus (the elderly, frail and infirm) are not the people who suffered most from the impact of lockdown (the young and many working adults).

There's no equity in any of this, but I was told again and again that segmentation and shielding should be resisted because it placed too much of the burden on a minority of the population, even though this was the same minority who were most at risk. I heard that argument caricatured as: everyone died, but at least no-one was saved unfairly.

The tragedy is that something like this actually happened. We did not do all we could to prevent a minority of the most vulnerable people getting infected with novel coronavirus because some influential voices thought suppressing the virus in the whole population was the *only* acceptable way forward.

The result was that people died who might have been saved if we'd put more effort into protecting them.

Improving awareness of the risks

The first step to doing more to protect those at greatest risk is to make everyone aware of where the risks actually lie. That's risk communication, and it's a vital task. We can only make good decisions – whether as individuals or as policy-makers on society's behalf – about how to respond to a pandemic (or any threat to our health and security) if we properly understand the risks.

I once wrote a report for the Food Standards Agency making the case that risk communication was all about exchanging information and must be open, honest, transparent and fully consistent with the scientific evidence available, acknowledging uncertainty and knowledge gaps. Other government departments and agencies have carried out similar exercises making similar recommendations. Unfortunately, risk communication during the pandemic failed to live up to these simple ideals.

The single most important thing to say about risk is that there is no such thing as zero risk. As individuals, we accept some level of risk in everything we do, whether it is travelling by plane or using a step-ladder. Public health scientists – though not, in my experience, all government ministers – understand this. Managing risk is a tough task because whatever you do some people will claim you are over-reacting and others will say you are not doing enough. That's because everyone has their own views on what constitutes sensible precaution and what is to be derided as interference by the nanny state.

For this reason, personal risk and public health have an uneasy relationship at the best of times. Sometimes government is content to issue advice: for example, we are advised not to drink too much alcohol but how much we actually drink is left up to us. Sometimes government is more prescriptive: the result is copious health and safety legislation.

Government tends to get most heavily involved when our actions have the potential to harm others and not just ourselves: road safety is a good example. With novel coronavirus our actions do have the potential to harm others, unknowingly, invisibly and often indirectly. Accordingly, the government took full charge and that influenced the way that the risks were communicated – sadly, not for the better.

As we've seen, from March 23rd onwards, the UK's response to novel coronavirus was centred on suppressing the virus, expressed as keeping the R number below one. The effort to suppress the virus – primarily by means of lockdown – was presented by the politicians as a community endeavour, which it was. To encourage collective action, the government appealed to a sense of community spirit, summed up as 'we're all in this together'. This approach was consistently supported by polling data showing that people were prepared to put their own freedoms on hold for the sake of everyone else.

However, there was the potential problem that – as we have seen – the actual risk to more than half the population was extremely low. Government advisors were concerned that this might reduce acceptance of lockdown. A solution to the problem was suggested by the SAGE subgroup SPI-B, which advises on behavioural science. Their advice dated March 22nd 2020 was first made public by BBC Television's *Newsnight* programme. The crucial section reads: *A substantial number of people still do not feel sufficiently personally threatened; it could be that they are reassured by the low death rate in their demographic group... The perceived level of personal threat needs to be increased among those who are complacent, using hard-hitting emotional messaging.*

With all due respect for my colleagues on SPI-B, this was

unwise. The government was contemplating introducing measures that were certain to cause a great deal of harm to a great many people who were *not* at great risk from novel coronavirus. SPI-B's advice gave them a charter to exaggerate the risks to ensure compliance. Michael Gove – Minister for the Cabinet Office – took full advantage, declaring that 'we are all at risk' and 'the virus does not discriminate'. The BBC News backed up this misperception by regularly reporting rare tragedies involving low-risk individuals as if they were the norm. No wonder there was such heightened anxiety in late March and early April 2020.

One of the problems with giving the impression that we're all equally at risk is that it dilutes the message to those who genuinely are at high risk of a severe outcome of infection. Talking to people over the course of that summer – including a House of Lords Select Committee in June – I realised that most of them were unaware of the huge variation in risk across the population. The list included not only politicians, journalists and teachers but many scientists and even some doctors.

The reality of novel coronavirus is that we need to be far more concerned about some people than others. As we have seen, 75% of deaths occur in the most vulnerable 5%. The more effectively we shield them from infection the more lives we save and the more we lessen the pressure on the NHS. If we fail to understand – or have been misinformed about – the increased risk to the elderly, frail and infirm then the less likely we are to take action to mitigate that risk.

I worry that this was part of the reason why we failed to protect care home residents and vulnerable people in the community properly. We weren't focused on the minority at greatest risk from novel coronavirus because too few people understood just how unequally distributed the risks were.

Fortunately, the government and the BBC are not the only

sources of information to help people gauge the risk to themselves and those around them. Over time, the message did start to get through that some were at far greater risk than others.

Even so, there was still a tendency to revert to the argument that we're all at risk every time the government felt the need to tighten restrictions. After all, for those arguing that the only way to tackle novel coronavirus is to suppress the virus, the notion that 'the virus does not discriminate' is helpful, even if it isn't accurate. It justifies lockdown much better than the reality that most of the risk falls upon a small minority.

CHAPTER 8

CHILDREN AND SCHOOLS

We have arrived at one of the most disappointing episodes in the UK's response to the novel coronavirus epidemic: the closing of schools. There was never any compelling evidence that school closures would have much public health benefit, but we did it anyway. To explain how this happened, I need to start with some recent history.

Schools and flu

The 2009–10 swine flu epidemic in the UK was driven by schools and schoolchildren. The virus arrived during the summer term and the epidemic grew rapidly until the beginning of the school holidays. The epidemic then declined until schools went back in the autumn, at which point it picked up again. School-aged children had the highest infection rates. Around seventy children died, with a case fatality rate of 0.02%.

There were two reasons why swine flu disproportionately affected children.

One is that many adults had a degree of cross-immunity having previously been infected with influenza strains similar to swine flu. Children are more prone to many respiratory viruses simply because they have had no previous exposure.

The second reason is that school-aged children tend to

make more contacts than any other age group, which gives them more opportunities to get infected. Typically, they have approximately 50% more contacts per day than their parents' age group and at least double those of their grandparents'.

Schools provide opportunities for children to mix, and also teachers, non-teaching staff and parents. Schools have another impact too: if the children are not at home it is easier for the parents to go to work. For these reasons, schools are seen as important hubs in the network of contacts that allows an infection to spread through a community. Secondary schools play a bigger role than primaries as they are mostly larger and have bigger catchment areas.

The epidemic models built by SPI-M over the years included all these features. In 2008, a large data set on contacts within and between different age groups was used to model a theoretical epidemic caused by a completely novel respiratory virus. The study predicted that school-aged children would be most affected, a result apparently validated by the swine flu pandemic the following year. That is why the early novel coronavirus models all indicated that schools would make a substantial contribution to the R number and school closures would be an effective way to slow the pandemic, perhaps the most effective single measure we could take.

This was the backdrop to the UK's preparations for the next pandemic. Novel coronavirus wasn't flu but it took time to overturn the preconception that schools and school-aged children would be central to the way the epidemic unfolded.

Evidence versus expectation
The first clues that novel coronavirus might be different from our flu-laden preconceptions come from the original SARS coronavirus. We knew that during the SARS epidemic in 2003 children suffered less severe symptoms and none died,

although some older children were hospitalised. We also knew that children with SARS were not highly infectious, even in school settings. One review of the evidence found only one instance of transmission by a child. We knew back in January 2020 that the novel coronavirus was a close relative of the SARS coronavirus; that should have made everyone wary about any preconceptions that schools would be important.

The early data from China supported the idea that the novel coronavirus was similar to SARS with respect to children. The disease was generally much milder in children than adults and there was a lower fatality rate even among symptomatic cases (and we now know that in children most infections are mild or asymptomatic). Contact tracing studies struggled to find any evidence of children passing on infection either. In June 2020, my team did a thorough search of the published literature and we only found eight reported instances of a child passing on infection from anywhere in the world. There had been a lot of contact tracing studies published by then so this was eight out of many thousands.

Only one of those eight was in a school, and we found it in a report of a study of five primary and ten high schools in New South Wales, Australia, that was published in late April 2020. In the fifteen schools in the study there were nine cases in students that were thought to have been acquired outside school. Just one secondary case was reported in a student – in one of the high schools – out of over five hundred students in the same class as a case. No staff members were infected by a student.

The New South Wales study does not give the impression of a virus that is easily transmitted in the school environment. The study captured all the key features of novel coronavirus transmission in schools and demonstrates that we had the evidence as early as April had we chosen to act on it.

Other data supported this assessment. No large outbreaks had been reported in boarding schools anywhere in the world, suggesting limited transmission even where children were living together. There was a steady trickle of scientific papers reporting that children were at low risk from novel coronavirus and that they were less likely to transmit the virus even if infected, describing the evidence in favour of school closures as 'equivocal' and warning that interventions aimed at schools may have 'a relatively small impact'.

Over the course of 2020 more research was published on the epidemiology of novel coronavirus in children, much of it based on further contact tracing studies. Several of these papers claimed to demonstrate that children were just as infectious as adults. On closer inspection, this turned out to be only half the story. It seems that children are just as infectious as adults *if* they are showing symptoms. The catch is that children rarely show symptoms.

There were occasional reports of children who did become severely ill with novel coronavirus – fewer than a hundred in the UK during the first wave, five of whom died. Though details are not routinely released on individual cases, where we do have information the children who died were, sad to say, already very poorly with underlying health conditions. It goes without saying that these are tragic cases, but they do not alter our understanding of the epidemiology of novel coronavirus.

We would like to know why healthy children are so unaffected by novel coronavirus. So far, no-one has come up with a convincing explanation. One idea is that children are protected by recent exposure to other coronaviruses which they experience as common colds. I'm not so sure.

Factors like previous exposure to an infection are termed 'environmental' to distinguish them from genetic traits. Severe illness in healthy children is so extraordinarily rare that it's difficult to see how environmental factors can be the explanation: it would have to apply to almost every child in the world, but not to every adult. It's more likely that whatever protects children must be hard-wired into the ageing process, perhaps to do with the ageing of the immune system, particularly around puberty.

Counterarguments

Despite the absence of any evidence that children were at significant risk, there was considerable resistance to the idea that it might be safe to keep schools open – as I've explained, it challenged firmly held preconceptions. I completely accept that a lot of the resistance was driven by a genuine and deeply felt concern for the safety of the children – I am, after all, a parent myself – and there was an understandable feeling that it was far, far better to be safe than sorry. That said, we still have to interpret the evidence as objectively and dispassionately as we can, or we may make poor decisions.

One important issue was that some children who had had novel coronavirus infections went on to develop MIS-C (multi-system inflammatory syndrome in children). The syndrome is rare – there had been fewer than forty reported cases in the UK by the end of April 2020 and though it can be serious no child had died.

Naturally, when these first MIS-C cases were reported this heightened concerns over children's safety and therefore the risk in schools, even though many children become infected with novel coronavirus at home where they come in closer and more prolonged contact with adults. Over time, these concerns at least partly abated, although MIS-C cases continued to accumulate during the winter of 2020–21.

Another concern was that Public Health England was reporting dozens of novel coronavirus outbreaks in schools during the summer term even though few children were in school at the time. These data were frequently cited by politicians as a reason to be cautious about re-opening schools.

You will recall that just two linked cases can be called an outbreak, so I suspected the problem was being exaggerated and asked my team to inspect the raw data in these reports closely. We found that not all the 'outbreaks' had any confirmed cases at all, just suspected ones. We also found that a large minority involved only staff, and most of the rest involved children who were likely to have acquired infection at home, leaving only a handful of cases acquired by children (or staff) in the school environment. There were cases in schools but these were playing a minor role in the epidemic and the risk to children remained extremely low.

Looking further afield, however, there were some large outbreaks in schools. There was one in New Zealand (apparently linked to a social event involving parents), one in France (which seemed to affect the entire community but was centred on a large school), and one in Israel (which seemed to involve a super-spreader event).

For me, these were the exceptions that prove the rule: the whole world was alerted to the possibility of large novel coronavirus outbreaks in schools and yet there was just a handful of examples. Nonetheless, though the risk is low it is not zero and that justifies some public health response, as long as it is a proportionate one.

Public Health England entered the debate again during the June outbreak in Leicester. This was widely reported – not least by Matt Hancock – as 'disproportionately' affecting children, a statement that caused a great deal of concern.

Again, my team took a close look at the report on the Leicester outbreak. It was true that cases were rising among school-aged children, but they were still considerably fewer than the cases in younger adults, and those were rising at least as quickly. There was no evidence that children were disproportionately affected in Leicester. The data were consistent with the by then typical pattern that if cases start to rise in adults there is spill-over into children, much of which is likely to occur in the home.

Another worry was that – even if children were at extremely low risk themselves – they might pick up an infection and pass it on to someone more vulnerable, such as an elderly grandparent or parent with an underlying health condition. It's a valid concern but, to decide just how we should react, we need to think about where the risk really lies. To do that, we need to consider the whole household. Some vulnerable adults will be the only adult in the household, but many will live in households with one or more other adults. Those adults are also a potential risk, perhaps more so than the children. In those circumstances, the argument for keeping children out of school is also an argument for keeping the rest of the household at home too.

This brings us back to the chain of trust. Every member of the household is part of that chain, children included. This kind of situation does need careful managing – we need to do all we can to protect vulnerable individuals – but there are ways to do that other than by closing schools.

The teaching unions also got involved in the debate. Their concerns seemed to reflect a misunderstanding of the risk to teachers. There was an entirely justified worry about teachers in the vulnerable category, and the same applied to any vulnerable person in the workplace. Overall, however, there was never any evidence that teachers were at elevated risk.

An Office for National Statistics survey published in November 2020 confirmed that teachers were no more likely to have had a novel coronavirus infection than members of any other profession. Primary school-teachers had the lowest levels of infection in the survey. When you think about it, these are telling observations. It's self-evident that teachers have a lot of contact with a lot of children – far more than most other people. If children are anywhere near as infectious as adults then you'd expect exceptionally high levels of infection in teachers. This isn't what the data show, which can only tell us that children in school are not that infectious.

The indirect contribution of opening schools to the R number was also widely cited as a reason to keep them closed. It's likely that there is a modest contribution but to this day it remains hard to find much sign of it in the case data. Before the start of the 2020–21 school year there was little direct evidence on this question anyway. The evidence we had came mostly from mathematical models, and the starting point for the models was that novel coronavirus is similar to influenza. With regard to schools, it isn't.

To recap, the arguments for closing schools were that the children were at some risk, the staff were at high risk and that schools would drive community transmission, thereby raising the R number. None of these concerns were supported by the epidemiological data and surely we should expect compelling evidence before we took such a serious step as closing schools. Sadly, this remained a minority view and none of the advisory committees were willing to recommend re-opening schools, so they stayed closed for the rest of the summer term.

Correcting misconceptions

In June 2020 I was asked by Moray House, the University of Edinburgh's School of Education, if I would give a virtual seminar on novel coronavirus in schools to a group of a few dozen head-teachers. I was happy to do so and presented much the same evidence that I have recounted here.

I was completely unprepared for the reaction I got. The

head-teachers looked astonished; most of what I said was completely new to them and contradicted what they had understood about the threat to the children and the staff. I'm told that the recording of the talk was widely circulated. Yet that shouldn't have been necessary; the same information should have been available through the Department for Education or other government channels.

Not long afterwards I contributed to a briefing to Scottish cabinet ministers on the risks of re-opening schools. Given the low incidence of novel coronavirus infections in Scotland at the time, together with what we now knew about the huge influence of age on the severity of infection, my team did some simple estimates of the risk of catching a fatal infection at school.

Our calculations put the risk to adults in the school at below one in four million per day, comparable with the risk of dying from an accident in the home. The risk to the children was too low to estimate properly but, at less than one a billion per day, was far less than the risk of being killed in a traffic accident on the way to and from school. In fact, it was of the same order as being struck by lightning. The ministers seemed no less astonished than the head-teachers had been. They had a different impression, but I never found out where it came from; it certainly wasn't from any epidemiological evidence.

I also got drawn into the debate on re-opening schools raging in the press. Both *The Times* and *The Mail on Sunday* made a great deal of a statement of mine that there was no evidence of any school-teacher ever being infected with novel coronavirus by a student in a classroom. The statement was accurate at the time but, as usual, the headline wasn't the whole story. I was not claiming that such a thing was impossible only that – given the intense scrutiny on schools – the lack of published examples suggested it wasn't common.

Again, it's about relative risk. The evidence is that if a school staff member gets an infection then it is at least as likely to have come from another staff member than a pupil (and, of course, staff can get infected out of school as well). There's a good argument that the most dangerous room in a school is not the classroom, it's the staffroom.

I got a trickle of hate mail every time I suggested that schools are not nearly as risky – for staff or students – as many people have been all too willing to assume. Most of my mail on this topic, however, was straightforward scepticism. Many parents, teachers and teachers' spouses (who seemed to be the most concerned of all) wrote to tell me that either I was misinter-preting the data or the data were wrong. Apparently, everyone 'knows' that children are walking virus factories so how could I be claiming that they aren't. Well, for flu and many other respiratory infections they are, but not for novel coronavirus.

Closing schools was not much help in controlling the epidemic, but it was hugely damaging and disruptive. During the summer, several reports were published setting out the harms being caused to children by denying them face-to-face schooling. The quality of online learning was patchy and students from lower-income households were likely to be at a greater disadvantage. In July, the Royal Society of London published a report – I mentioned it in Chapter 5 – citing harm to the children's mental and physical health and to their long-term prospects, as well as significant long-term damage to the economy through having a less well-educated workforce.

The Royal Society report helped to turn the tide and the scientific consensus shifted quite quickly. Russell Viner – President of the Royal College of Paediatrics and Child Health and SAGE member – was an influential and vocal advocate in favour of re-opening schools from the outset. International experience added to the evidence: Sweden didn't close schools for under fifteens at all, but still brought its epidemic under control. Denmark re-opened its primary schools in April and secondary schools in May, with few problems.

Re-opening schools and its impact

The groundswell of evidence that schools were safe and keeping them closed was harming the children won the day. Ministers changed course and schools re-opened in full in Scotland in August and in England and Wales in September.

To my mind, this was one of the biggest successes of evidence-based policy making during the pandemic and I'm proud of my team's contribution to making that happen. The irony is that this was a triumph of science over scientists. There was never any evidence that children were at significant risk from novel coronavirus or that schools were driving the epidemic in the UK. The only 'evidence' was scientific opinion based on knowledge of pandemic influenza, a different infection. Yet schools across the UK were closed (except to a small minority of children) for more than three months in the first half of 2020, affecting around ten million children and their families.

When schools did re-open there were – as expected – minimal repercussions for public health. There were cases in children but serious illness remained extremely rare. There was also a rush of demand (naïvely unforeseen by the testing system) for testing of children with possible symptoms. Positive cases in children did rise in the autumn, though they rose faster in other groups, notably university students.

In schools, the rise in cases was most apparent among older teenagers. Some epidemiologists were concerned that this age group had a higher prevalence of infection than their parents' age group. Older teenagers are considerably more susceptible and more infectious than younger children, but not more than adults. The obvious explanation for the pattern we were seeing was that these children were living lives far closer to pre-pandemic normality than any other age group, so their numbers of contacts were closer to normal too. Even so, sixteen-, seventeen- and eighteen-year-olds make up only a small fraction of the population so it was hardly likely that they were driving a resurgence of the epidemic that was affecting everyone.

This is not to say the pandemic didn't cause problems in schools in 2020. It certainly did in some, though the majority of schools had few or no cases. Where there were outbreaks these usually involved staff, sometimes bypassing the students altogether. Any disarray was due more to the need for children and staff in the same class or bubble to self-isolate, some of them on multiple occasions during the course of the autumn term. We had anticipated this in the Scottish Advisory Group, warning before term started that schools needed to plan the steps they would take if cases did occur so as to minimise disruption.

I accept that it would have been difficult not to have closed schools when the UK locked down in March 2020 – if only as a precaution – but it took far too long to realise that there was no good reason to keep them closed. Some argued that it was too risky to re-open schools while levels of infection remained high. Yet we knew at the time that the main risk from novel coronavirus is to the elderly, frail and infirm, not to schoolchildren. If we'd re-opened schools – perhaps for all but the oldest children – there would have been limited impact on the epidemic and we would have avoided massive disruption to the education of millions.

CHAPTER 9

TEST, TEST, TEST

Testing could and should have made much more a difference than it did during the first year of the novel coronavirus pandemic. Unfortunately, there was a great deal of confusion about the role of testing, which led to two seemingly contradictory problems. First, there was an overpromising of what the testing being done would achieve. Second, there was an underappreciation of what testing could achieve, if we put in enough effort and investment.

The role of testing

This confusion was not helped by the well-publicised call to 'test, test, test' made by the Director General of the World Health Organization in March 2020. He used the phrase in the context of trying to contain the virus through contact tracing, a strategy which ultimately proved insufficient in most regions of the world, including the UK. Ironically, I completely agree with the World Health Organization's emphasis on testing, but without clarity on what testing could and couldn't achieve the exhortation to test, test, test muddled rather than clarified how we should tackle novel coronavirus.

Testing can benefit the person tested in several ways. A positive test result confirms a clinical diagnosis and may bring about the diagnosis if the patient presents with unusual symptoms. It also affects how the patient is managed within the health care system. A negative test result may allow a person to return to work or be released from self-isolation or quarantine.

I will focus mainly on another kind of benefit: how testing can be used to reduce the transmission rate, which is a benefit not to those tested but the people they come in contact with. That said, by itself, testing does nothing to stop the virus spreading. That is achieved by the self-isolation and contact tracing triggered by a positive test result.

As we have seen, in February and early March 2020 the UK's strategy was containment, meaning that when outbreaks or individual cases were suspected they were quickly followed up and those affected told to self-isolate. Contact tracing was then carried out for cases that tested positive.

The containment strategy aimed to limit the number of onward transmissions so that the R number was held below one and the epidemic could not take off. This was always optimistic – as we saw in Chapter 6, self-isolation and contact tracing alone were unlikely to be enough to keep the epidemic under control – but the strategy could help to reduce the transmission rate as long as case numbers remain low enough that the contact tracers could keep up.

Unfortunately, case numbers did not remain low and the UK's limited testing capacity quickly became overstretched. Matt Hancock gave a nice explanation of why this happened at a House of Commons Select Committee in November 2020. He pointed out that while test capacity was growing linearly (by a fixed amount every day) the epidemic was growing exponentially (by the same ratio every day).

This is a familiar line of argument that dates back to 1798 when Thomas Malthus made the same point about food supply, population growth and the inevitability of overpopulation.

Malthus's idea was controversial among economists but is said to have had a profound influence on Charles Darwin and his theory of natural selection. I never expected it to crop up in a discussion of diagnostic capacity, but it was a valid point for the Health Secretary to make. Sooner or later, exponential growth beats linear growth, and so it proved.

In March 2020 the testing of suspected novel coronavirus cases in the community was abandoned in order to concentrate on testing in hospitals. As testing capacity increased over the following months it became possible to re-introduce testing in the wider community. There were many reasons why this was important to do. I shall discuss these under the headings get-back-to-work tests, case finding, surveillance, fit-to tests and test-on-request, and mass testing.

Some background on diagnostics

First, though, we need some background on how testing works, starting with the different types of tests and what they tell us. There are two broad categories. Tests for the presence of the virus tell you if you are currently infected. Tests for the presence of antibodies to the virus tell you if you have been infected with the virus in the past.

The tests for the presence of virus used most widely in 2020 are based on a technique called RT-PCR. An RT-PCR test targets short sections of the virus genome. If the correct genome sequence is detected in a nose or throat swab then the test result comes back positive. Several RT-PCR tests for novel coronavirus were developed by labs around the world within two or three weeks of the virus genome being published on January 11th.

There is another type of test for the presence of virus called an antigen test. An antigen is a protein molecule on the surface of the virus. The antigen elicits an immune response called a specific antibody response in an infected person or (for the purpose of producing a test kit) an infected rabbit or mouse. An antigen test is based on a technique called ELISA that uses these specific antibodies to detect the antigen. A nose or throat swab will give a positive test result if it has picked up viral antigens that are recognised by the antibodies.

Then there is an antibody test. This too is based on ELISA – it's like the antigen test but the other way around. The test kit contains viral antigens. If antibodies to the virus are present in a tiny blood sample then they will bind to these antigens. If the test result comes back positive it means that you have previously been infected, possibly many months ago.

We also need to know how accurate the tests are. There are two components to accuracy, sensitivity and specificity. Sensitivity is the fraction of positive cases that test positive. No test is perfect so some people who have the infection will test negative; these are called false negatives. The sensitivity of the RT-PCR test for novel coronavirus can be calculated in different ways. For true cases with classic symptoms it is probably not far off 100%. False negatives are much more likely early in the pre-symptomatic phase when the level of virus is low. That's a problem: it means that we cannot tell if people have been infected very recently. Conversely, in the late stages of an infection the RT-PCR test may detect virus fragments even if there is no viable virus present. This isn't, strictly speaking, an error – there is evidence of recent infection there that the test detects – but it implies that just because someone tests positive they are not necessarily infectious.

The other important metric is specificity, the fraction of suspect cases who are uninfected and test negative. People without the infection who test positive are called false positives. The RT-PCR test has high specificity – meaning that it generates few false positives – because it targets unique sections of the virus genome. The false positive rate is estimated at less than 0.5%.

Overall, the RT-PCR test performs well; it has both high sensitivity and high specificity. It isn't perfect – no diagnostic test is, if only because human error is always possible. There is plenty of scope for human error during testing – such as taking a poor swab or contamination of the sample in the laboratory – so quality control is vital.

The World Health Organization introduced another metric during the novel coronavirus pandemic: test positivity rate, the fraction of tests carried out that give a positive result. They advised that if more than 5% of test results are positive then an epidemic is not under control.

Oddly, the World Health Organization never explained where the 5% figure came from. There is no obvious scientific basis for it. All else equal, a lower rate is preferable to a higher rate, and an increasing trend is a concern, but that's as far as I'd go. Despite this, the Scottish government put great store by the 5% figure, even though they didn't calculate it quite the way the World Health Organization prescribed. This matters because it influenced consequential decisions such as whether to impose or lift restrictions, decisions that should not be based on arbitrary indicators.

The symptoms of novel coronavirus infection are not specific – they could be caused by other respiratory infections, or even allergies – and so many suspected cases will turn out be false alarms. This situation creates a big demand for testing to establish that a suspect case can cease to self-isolate. I call these 'get-back-to-work' tests as they were much in demand for that reason by the police and other services during the early stages of the pandemic.

When testing capacity is limited – as it was in the UK in early March 2020 – get-back-to-work tests are low priority because if the person is self-isolating effectively then the job of preventing transmission is already done. It would be better to use the remaining test capacity in an effort to reduce spread in high-risk settings such as hospitals. There was also a concern that false negative results – though rare – could mean that people are allowed back to work in error. On the other hand, there's a big cost to not deploying get-back-to-work tests if the majority – perhaps the great majority – of suspected cases do not have novel coronavirus after all and so do not need to self-isolate.

The situation changed as testing capacity grew and contact tracing became routine. As we saw in Chapter 6, it is vital to test all suspect cases as quickly as possible so their contacts can be traced. This has the added – though incidental – benefit of allowing those testing negative to go back to work. Unfortunately, that benefit is more than outweighed by a new problem. Tracing simply moves the get-back-to-work issue one link further along the chain of transmission. Now the contacts – usually more than one per case who tests positive – are trapped in self-isolation with no automatic right to testing, unless they in turn develop symptoms.

The idea of testing to release contacts of cases from self-isolation was strongly resisted. This was illogical; if a negative get-back-to-work test releases someone with apparent symptoms from self-isolation there is no reason why the same cannot be done for contacts of cases. One caveat is that the RT-PCR test may miss low levels of virus during the earliest stages of infection. This is easily dealt with by waiting for a few days. Exactly the same argument applies to anyone asked to quarantine for any reason. Later, I'll take this argument one step further and ask whether testing could be used to do away with self-isolation of contacts and quarantine altogether.

Increasing testing capacity became a government priority and throughout April 2020 there was tremendous media interest in whether the UK government would meet Matt Hancock's self-imposed target of one hundred thousand novel coronavirus tests per day. With a little massaging of the numbers they did. My colleagues who had been repeating the 'test, test, test' mantra had got their wish.

How test and trace works

At the end of May, the UK launched NHS Test & Trace, more usually referred to as Test, Trace and Isolate or TTI. In Scotland the system was called Test & Protect. TTI was responsible for increasing the volume of testing and delivering contact tracing and outbreak investigation. Twenty-two billion pounds was committed to the first twelve months of TTI, and there was plenty of interest in how well it performed. So how did it do?

Well, though TTI was widely derided for falling short of the 'world-beating' system that the government promised it was, after a shaky start, quietly successful. Testing capacity grew steadily and reached over half a million per day by the end of 2020, with over 90% of test results received the next day. The performance of contact tracing improved steadily such that well over the target 80% of contacts were being located and asked to self-isolate, almost all of them within twenty-four hours. These performance statistics didn't spare TTI from a damning House of Commons Public Accounts Committee report in March 2021 that argued that it had made barely any difference to the progress of the epidemic.

In TTI's defence, it has two built-in weaknesses, both entirely beyond its control.

First, it relies on people suspecting they might be infected and voluntarily coming forward for testing. If they do not then TTI is powerless. We'll look at the extent of this problem and possible solutions a little later.

The second weak link is whether or not people self-isolate

when asked to do so. As the behavioural studies had shown, they often did not. Even if the contact tracing is going perfectly, there is only so much TTI can achieve if cases are not being found or are not self-isolating when they are.

Not everyone agreed with that assessment. There was no shortage of advisors and commentators insisting that the main reason the UK suffered such a severe first wave was that we'd been too slow to develop and deploy testing capacity. Other countries – it was claimed – did better because they were more serious about testing.

This argument suffered a blow when the UK's massive investment in TTI failed to prevent case numbers rising in the autumn of 2020 and reaching unsustainably high levels at the end of the year, so that we ended up back in lockdown. That wasn't what those who had been advocating 'test, test, test' were expecting to see: TTI was supposed to be our way out of the pandemic.

The critics' response to this turn of events wasn't a very scientific one: the goalposts were moved and it was claimed that TTI would have prevented the second wave if only we were doing it better. That claim doesn't stand up to scrutiny. By the time the second wave arrived every country in western Europe – just like the UK – had far greater testing capacity than they did at the start of the pandemic. The investment in testing was enormous but it didn't save them. Just like the UK, all of those countries suffered a second wave at least as bad as the first. The lesson is clear: TTI is just one element of an effective response to novel coronavirus – it was never the magic bullet that some had promised.

The importance of surveillance

Testing has another, quite different, role too. It is central to effective surveillance of novel coronavirus cases. Surveillance is defined in the 2005 International Health Regulations as *the systematic, ongoing collection, collation and analysis of data for public health purposes and the timely dissemination of public health information for assessment and public health response as necessary* and is regarded as a critical component of epidemic management. I had stressed the importance of setting up a novel coronavirus surveillance system as quickly as possible in my January 21st e-mail to the CMO Scotland.

It is possible to do surveillance for respiratory diseases based on symptoms – public health labs routinely report the incidence of ILIs (influenza-like illnesses) on this basis. Symptom-based surveillance is handicapped by low specificity but it can still be useful, both for influenza and novel coronavirus. An excellent app-based system called ZOE was set up in the UK as early as March 2020, reporting trends and analysis of surveillance data for novel coronavirus symptoms throughout the epidemic.

Two surveillance programmes based on virus tests were set up later in the spring, one by the Office for National Statistics and one – called REACT – by Imperial College London. The weekly results of both were eagerly awaited during the autumn second wave.

Each survey in these programmes involved sending test kits to tens of thousands of people of whom typically about one-third would return a sample. The proportion testing positive in these surveys is a measure of the prevalence of infection and allows the number of people with the virus at any one time to be estimated. There is some scope for bias – for example if people with symptoms are more likely to return a sample – but this can be accounted for to some extent and these surveys were regarded as reasonably reliable, even though the estimates they gave could differ by as much as a factor of two.

One of the most important findings from the Office for National Statistics and REACT surveys was that they consistently estimated prevalence at up to five times higher than you'd expect given the numbers of positive cases reported. We have to allow for pre-symptomatic infections that would be reported subsequently, but the implication is that many infections – possibly more than half – were being missed, even towards the end of 2020 when testing capacity was much greater.

We cannot know whether those unreported cases self-isolated – I suspect many did not – but we can be sure that their contacts were not traced. I thought this was a huge problem – I likened it to fighting the pandemic with one hand tied behind our backs – but it attracted little attention.

Both the Office for National Statistics and REACT also carried out antibody surveillance. In the summer of 2020 the antibody surveys indicated that around 5% of the population had been exposed, rising to almost 10% in London, where the epidemic had been about a week ahead of the rest of the country prior to lockdown on March 23rd. That works out at over three million people.

This was a useful number to know. It provided a denominator for calculating the infection fatality rate – which is how we know it was about 1% during the first wave. It told us that we were still a long way short of levels of exposure needed for herd immunity to play a significant role. It provided further evidence of how many cases had been missed, at least nine out of ten during the first wave given that only three hundred thousand people had tested positive by the end of July.

Screening as a public health tool

As testing capacity grew it became feasible to test many more people for pre-symptomatic or asymptomatic infections, in addition to the routine testing of suspect cases. This 'blind' testing is referred to as screening.

During the summer, a Scottish advisory subgroup led by David Crossman did a lot of work trying to prioritise groups for testing. The first few categories were agreed fairly quickly: contacts of cases, health care workers and care home workers. After that it became more difficult.

The best choice of priorities depends on your rationale for testing. If we are simply trying to find as many infections as possible then the optimal strategy is to target groups with high prevalence. The problem with that is there are only modest differences – no more than two-fold – between demographic groups or between occupations. There can be bigger differences between localities but these fluctuate over time. This could lead to prioritising taxi drivers in Aberdeen one week and students in Edinburgh the next.

Another strategy is to prioritise testing where the consequences of the person being infected would be most serious. This means everyone in regular contact with people in the vulnerable category. This list would include members of the same household as a vulnerable person, informal carers and care-in-the-home workers – these are the 'shielders' as I defined them in Chapter 7.

I saw this as an extension of the same logic for prioritising health care workers and care home workers: to protect their vulnerable contacts. I strongly favoured such a strategy as it would contribute directly to reducing the toll of serious illness and death, but wasn't put into practice until much later in the epidemic.

The NHS was not the only source of testing, however. Various institutions and organisations were sourcing tests privately. An early adopter of this approach was the English and Scottish Premier Leagues who managed to complete the 2019–20 football season – albeit without fans in the stadiums – by virtue of regular testing of players, coaches and staff. For no good reason that I could see, few other organisations did the same at the time.

A similar logic extends to individuals too. There are many activities people might be prepared to do, be more confident in doing, or be allowed to do if they had access to testing. One of these was made available: the fit to fly test, a sensible precaution. I wanted to see other 'fit-to' tests: fit to visit a care home test; fit to stay with grandparents test; fit to go to church test; fit to attend a family party test; fit to go to a concert test; perhaps even a general fit to go to work test.

One of my colleagues dubbed this approach as 'test-on-request'. The demand was there, as we saw from the queues when the retailer Boots introduced fifteen minute in-store tests in October 2020. There was lukewarm support, however, from public health officials who weren't eager for testing to happen outside their jurisdiction. This attitude was unhelpful, but there was another problem with test-on-request: scale.

Pros and cons of mass testing

I mentioned the intense interest in Matt Hancock's self-imposed target earlier, but reaching one hundred thousand tests per day wasn't a pivotal moment in the course of the epidemic.

First, as we saw earlier, the public health benefit of these tests depends on who is being tested, why, how quickly the test results are received, and what actions follow from the test results. The narrow focus on numbers of tests risked side-lining these other vital considerations.

Second, one hundred thousand tests were nowhere near enough to have the kind of impact that would get us out of the first lockdown and spare us from future ones.

As far back as the second Scottish Advisory Group meeting in March 2020, I'd argued that we were going to need to test people on an unprecedented scale, perhaps even weekly testing of the bulk of the population. My reasoning was that if there was significant transmission during the pre-symptomatic phase of infection and by asymptomatic cases then there was little chance of keeping the virus in check by testing only sympto-matic cases and their contacts.

When I made that suggestion a senior government official gently admonished me by pointing out that the group's advice had to be realistic. Given that testing capacity was still limited at the time, it was a fair point, though only if you accept that back in March 2020 few (if any) in government fully under-stood how big a crisis this was going to be. My less gentle reply was that we were going to have to re-think what was 'realistic'.

The biggest barrier to mass testing in the early stages of the pandemic was the nature of the test. RT-PCR is not ideal for mass testing: it is slow, cumbersome and expensive. My virologist colleague Paul Kellam and I spent some time in April sketching out what mass testing using RT-PCR might look like. There would need to be portable laboratories at the test site where trained staff could process the tests immediately. The big advantage of testing on site was that it cut out travel times, often the biggest delay in getting test results quickly. In principle, it ought to be possible to get the result the same day. In practice, the logistics of doing this on a large scale across the whole country would be challenging to say the least.

The main requirements of a mass testing diagnostic are that it is easy to use, cheap, quick and accurate (meaning that it has high sensitivity and high specificity). RT-PCR meets only the last of these criteria: we needed something else. Ideally, we wanted what is known as a point-of-care test, meaning that the sample does not have to be transported to a laboratory for processing.

Several point-of-care tests were developed during the first nine months of the pandemic. The Boots in-store test used a small machine to process samples. Even easier is the now familiar lateral flow test for virus antigen – this resembles a home pregnancy test, and it too can be performed in the home. This test requires a nose or throat swab, though it can be used on saliva instead. As well as being easy to use it is cheap and quick – results are available within thirty minutes.

A potential stumbling block for these new tests was accuracy. I spent many hours in meetings with public health colleagues who felt both that the new tests were not accurate enough and that people could not be trusted to use them properly. There seemed to be an element of not wanting to lose control over testing in these objections, but there were more legitimate concerns too.

Part of the problem comes back to scale. Let's imagine that you are testing a million people and expect that around 1% – that is, ten thousand – are infected. For illustration, let's say that the test you are using is 95% sensitive and 99% specific. 95% sensitive implies that you will miss 5% of the ten thousand cases – that's five hundred false negatives. 99% specific implies that you will misclassify 1% of the nine hundred and ninety thousand people who are not infected – that's almost ten thousand false positives.

This is nothing new. Regulators require that the sensitivity and specificity of any new diagnostic test are formally evaluated before it is widely used so that we know how big the problems of false positives and false negatives are likely to be.

There are ways around both problems. The main concern about the five hundred false negatives is that they might consider themselves 'safe' to undertake activities – such as visiting an elderly relative or singing in a choir – where the consequences of being infected could be serious.

The obvious answer is not to undertake these kind of high-risk activities until you have tested negative twice over the previous three or four days. Even this can never be 100% fool-proof, but it gets closer. The concern about the five thousand false positives is that they would be required to self-isolate and their contacts traced, a huge waste of time and resources.

My public health colleagues were greatly concerned about the false positive problem, often presenting it as an insurmountable barrier to mass testing. They also saw it as a barrier to regular testing of anybody – particularly health care workers – on the grounds that they were bound to receive a false positive test eventually and have to self-isolate unnecessarily.

I didn't see why this was any different to diagnosis based on symptoms. The symptoms of Covid-19 are not specific, so self-isolation is recommended while awaiting a confirmatory RT-PCR test. That may take an additional forty-eight hours, but if the test result is confirmed then two days of transmission have been prevented; and if the result is overturned then two days wasted is far less problematic than ten, the standard requirement for self-isolation of a confirmed case. A similar strategy could be used to minimise disruption due to false positives thrown up by a mass testing programme.

The false positive problem is actually subtler than I have just described. It is a version of a well-known paradox called the Prosecutor's Fallacy. In our context, the person with a positive test result risks being falsely 'convicted' of having novel corona-virus. After all, the prosecutor can point to the test being 95% sensitive and 99% specific (the values I used earlier), which may well sound accurate enough to convince a jury. If the infection is common then there will be many fewer false positives than true positives, so the prosecutor is probably right.

However, the argument does not work nearly as well if the infection is rare. A prevalence of 1% – the illustrative value I used above and representative of the second wave in the UK – gives one false positive for every true positive. In legal parlance, there is now ample cause for reasonable doubt. For a prevalence of 0.01% – around the minimum value seen between waves in the UK – the ratio of false positives to true positives rises to almost a hundred to one. To put this another way, in that scenario more than 99% of positive test results come from people who do *not* have the infection.

This illustration should make it clear that false positives are truly a problem, one that some consider a fatal flaw of mass testing. Others – myself included – consider it a necessary inconvenience that can be dealt with by confirmatory testing. Mass testing will never be perfect but a few thousand false positives are surely less disruptive than locking down millions.

In August, Matt Hancock announced the Moonshot programme, population-wide mass testing. I loved the word 'Moonshot'. When I'd been asked early on in the pandemic how I would avoid lockdown I answered that one way was mass testing on a scale that sounds like science fiction, so 'Moonshot' was perfect. Back in August 2020 the idea of testing everyone in the country still looked a long way out of reach, but happily this piece of science fiction was destined to become reality.

The game-changing innovation was the new point-of-care tests, particularly the lateral flow test. Mass testing using this technology was piloted in Liverpool in early November. Liverpool was a novel coronavirus hotspot at the time and testing was offered to everyone at facilities set up across the city. Ideally, every person would be tested twice. Only around a quarter of the population took up the offer – less than that in some areas – but the survey did detect several hundred positive cases.

The low uptake was hardly surprising. The consequence of testing positive was having to self-isolate for ten days, inconvenient for anyone and impractical for many. The only – though important – benefit was knowing that you were a risk to others. The test-on-request approach would surely work better if it enabled people to do safely something they might not otherwise do (or be allowed to do) at all, but this wouldn't become mainstream thinking until well into 2021.

Despite the low take-up and reported problems with poor test sensitivity, the mass testing pilot in Liverpool was deemed by the UK government to have been a success and mass testing was rolled out to other areas later the same month.

Now we had another tool for keeping novel coronavirus under control. Point-of-care tests could be used to detect large numbers of pre-symptomatic and asymptomatic cases, helping to locate those many undeclared cases that were helping to fuel the epidemic. We were also on the way to the another important intervention, regular testing of shielders. If mass testing had been given higher priority then all this could have been achieved sooner. I had said on the Radio 4 news back in August that if the government was serious about mass testing then it needed to get a move on. Still, credit where credit is due, within nine months of the UK's first novel coronavirus cases a mass testing capability was duly delivered.

CHAPTER 10

LIVING WITH THE VIRUS

By July 2020, after more than three months of lockdown, numbers of novel coronavirus cases and deaths throughout the UK had fallen to far lower levels than in the spring. They remained low for the rest of the summer: Scotland recorded some days with no deaths at all, down from a peak of over three hundred a week in April.

The R number had been monitored closely from early March when it had a value of around three. R fell to around 0.7 during lockdown, meaning that every index case was infecting an average of well under one other, the epidemic was in decline and there was no prospect of a large-scale resurgence, though there was still the possibility of self-limiting outbreaks, perhaps linked to super-spreading events.

Beyond lockdown

Keeping the R number below one was now a policy objective, which meant that there was limited scope for easing restrictions, particularly if we continued to rely on social distancing to reduce the transmission rate. However, easing restrictions was a policy objective as well, which meant that there were two conflicting policies in play at the same time. This did not look promising.

There were some alternative, but more radical, options. 'Let-it-rip' – meaning not to suppress the virus at all so as to expedite the build-up of herd immunity – was not on the table in the summer of 2020, herd immunity strategies having been dismissed back in March, but it did re-appear in the autumn so we will come back to it later. The antithesis of let-it-rip was elimination, and that was firmly on the table.

Elimination means local eradication, pragmatically defined as zero prevalence of infection in a defined region for some agreed length of time. It requires an aggressive control programme to eliminate existing cases, a strictly enforced cordon sanitaire to prevent new ones coming in, plus a standing capacity to respond effectively if (more likely when) new cases do get in.

I knew what a disease elimination strategy looked like because I had previously helped to deliver one. This was in the context of the foot-and-mouth disease epidemic in the UK in 2001 when I was part of an ad hoc group advising David King, the CSA at the time.

Foot-and-mouth was a completely different problem – a livestock disease spreading between farms with no risk to humans – but it's an instructive case study because elimination was the unambiguous goal and it was energetically pursued. There was plenty of debate about the means – a livestock-culling programme – but no disagreement about what we were trying to achieve. The UK was declared officially free of foot-and-mouth disease less than a year after the onset of the largest epidemic in our history. Looking back now, that was an impressive achievement.

Eliminating novel coronavirus was a quite different proposition. Imprecise terms such as 'near elimination' and 'quasi-elimination' were quickly invented to reflect the all-too-real uncertainty as to whether elimination was a realistic option for Scotland. It was not.

The problem was the number of cases. Though relatively few cases were being detected in Scotland during the summer, this was a time when testing capacity was still fairly limited. The best estimates from SPI-M were that the incidence of new cases in Scotland was about ten times higher than was being reported. These estimates were validated when testing capacity grew rapidly in August and many more cases were detected, particularly among young adults who mostly have mild symptoms or no symptoms at all. When Office for National Statistics prevalence surveys started in Scotland in October 2020 their data agreed with SPI-M's estimates and not the Scottish government's daily case reports. The best estimate is that at no time in the summer of 2020 were there fewer than five hundred cases in Scotland, mostly undetected.

Of course, in theory five hundred cases could be reduced to a much small number, but if most of those cases were not visible to the contract tracers then the only way to do so would be to suppress transmission – that is, keep the R number as low as possible – by imposing restrictions on everybody.

There were two problems with this. First, it was contrary to Scottish government policy of relaxing restrictions, which could only increase the R number. Second, it would take a long time.

The second of these points was not fully appreciated because exponential decay is even less well understood than exponential growth. During the early summer of 2020 the number of cases was halving every two weeks. If you start from five hundred cases that means it would take about three months to get the number down to single figures. That's a lot longer time to be in full lockdown – and even then the infection hasn't been eliminated – but there's an extra twist. It's like exponential growth but in reverse. With exponential growth half the cases appear right at the end. With exponential decay half the cases disappear right at the start. We would get from five hundred to two hundred and fifty in the first two weeks – that's half the public health benefit. It takes months to get rid of the other half, with every successive fortnight inflicting the same hardship as the previous one for only half the gain.

For this reason, a greatly extended lockdown would be hard to justify as a public health intervention. The public health burden due to novel coronavirus in Scotland was already extremely low, the virus was far less of a problem than the restrictions in place to control it and prolonging the restrictions could only add to the harm they were causing.

There was a brief discussion of elimination for the whole of the UK, which would be even more challenging. My colleagues with expertise in the elimination of Ebola from West Africa and the near-eradication of polio worldwide felt that eliminating novel coronavirus from the UK would require at least a year of strict lockdown and might not be achievable even then. The countries – like New Zealand and Australia – who came closest to eliminating novel coronavirus in 2020 were the ones where the virus had never properly established. It was firmly established in the UK by the summer of 2020.

Another challenge was border controls, an essential element of an elimination strategy. Even New Zealand and Australia were having to contend with a steady trickle of outbreaks that had to be dealt with by localised lockdowns. If those isolated nations could not stem the flow of imported cases entirely then it was hard to see how the UK could, let alone Scotland by itself – there would be too much cross-border traffic, even with an outright ban on all non-essential travel.

My advice to Scottish government was that an elimination strategy for Scotland would surely be seen to fail. The idea was quietly dropped, though it was to re-surface the following year under a new name, Zero Covid.

Debate about travel restrictions

Even if elimination wasn't a practical objective, the discussion did raise the question of whether the UK – or Scotland – should introduce much stricter border controls anyway. This, it was claimed, would make the novel coronavirus epidemic easier to manage. It was never that straightforward but it's worth looking at the arguments.

There is no question that travel can spread an infection like novel coronavirus – that's how it arrived in the UK in the first place. Restrictions on travel are – in some circumstances – an appropriate and effective public health intervention, if a costly one. The key phrase is 'in some circumstances'. To understand the potential contribution of border controls and travel bans to suppressing the virus we need to think about when they are effective and when they are not.

A strict ban on travel may be appropriate when an infection is confined to a restricted geographical area. In this scenario, the aim is to keep the infection in. This was the situation in Wuhan in early January 2020 and was the basis of China putting a cordon sanitaire around the city with tight controls over movements in and out. At the same time the city was put into strict lockdown to try to eliminate the virus within. Unfortunately, novel coronavirus had already spread beyond Wuhan so these interventions came too late.

A ban on travel may also be appropriate in the mirror-image scenario, when the infection is widespread but absent from a restricted geographical area. The aim now is to keep the infection out. This kind of intervention has been used for centuries, as it was during plague epidemics in medieval times.

This was the strategy pursued by New Zealand and Australia. It only makes sense, however, if infection has been eliminated within the enclave. As we have just seen, for the majority of countries in the world – Scotland included – there was no realistic prospect of eliminating the virus in 2020. The best time to have introduced a ban on travel to the UK would have been in February 2020, before novel coronavirus was firmly established here. That could have had an impact, far more so than the same action taken once the epidemic was well under way.

Travel restrictions are expected to have much less impact if infection is already present in both source and destination. There are two scenarios: R below one and R above one.

In the scenario where R is below one there's a common misperception that imported cases can somehow 'tip the balance' and cause the epidemic to take off again. That's not correct. Unless the imported virus is more transmissible (we'll come back to that possibility later), if the R number is below one then none of the viruses – imported or not – will take off. Here's another way to put it: opening or closing the border has no effect on the R number.

I can back up this argument with a case study from within the UK. In the late summer of 2020 there was an influx of visitors to the Scottish Highlands & Islands, many from England (and many from other parts of Scotland too). Some decried this as foolhardy and recommended banning visitors to prevent a public health disaster. That would have been a huge blow to a regional economy that relies heavily on tourism and, having endured months of lockdown, was in dire need of a fillip. I argued that banning English tourists at that time would have minimal public health benefit. Even if cases were imported and even if the infection was occasionally passed on this would not spark a new epidemic because the R number in the region was well below one. In the event, tourism was allowed to continue and there were indeed few novel coronavirus cases in the Highlands & Islands, and even fewer linked to tourists.

You might think it is different if the R number is greater than one and incidence is rising exponentially. It is easy to imagine that importing cases now is adding fuel to the fire. That's not the right analogy though. In those circumstances imported cases are more like dropping lighted matches into the flames – they flicker briefly but make hardly any difference in the long run.

This is another illustration of Malthus's insight that exponential growth beats linear growth. For this reason, the fuel-on-the-fire argument works best in the earliest stages of an epidemic. A large influx of imported cases early on gives the epidemic a head start, instantly generating an increase in incidence equivalent to multiple doublings. Super-spreading events in the early stages can have the same effect. The epidemic would still happen without the influx or the super-spreading event, but it could be delayed by many weeks.

It's worth mentioning that travel bans (internal or international) can also have an impact on the epidemic for a different reason. We often travel to socialise or participate in other activities involving contacts with others. If banning travel means these activities don't happen (which depends on what people do instead) then it reduces the overall contact rate and lowers the R number. Viewed this way, a travel ban is an element of social distancing designed to suppress the virus.

The general principle remains that border controls and travel bans have greatest impact when they prevent the movement of infection from a location where it is already established to a location where it is absent. This assessment was supported by a study published in late 2020 confirming that – for most countries in the world – imported cases had made only a small or very small contribution to their novel coronavirus epidemic. Not none though. So there is an argument for more proportionate interventions than banning cross-border travel entirely. There are two options: quarantine and testing.

Quarantine and how to avoid it

Quarantine is another ancient public health intervention. Quarantine works by preventing someone deemed at high risk of being infected – though not necessarily showing symptoms – from infecting others. Self-isolation of novel coronavirus cases and their contacts is a form of quarantine, although I shall continue to refer to these measures as self-isolation and reserve the term 'quarantine' for people who are not cases or contacts.

A crucial question is the level of risk that triggers the requirement to quarantine. We can gauge an appropriate level by referencing other categories of risk, such as a contact of a confirmed case required to self-isolate. From surveillance studies in the autumn of 2020, we now know that roughly one in thirty, or 3%, of contacts tested positive. That doesn't sound high but it was several times higher than the level of infection in the general population at the same time.

So contacts are at relatively high risk, as you'd expect, but after that it gets much more difficult. In principle, anyone who has engaged in a high-risk activity or has been to a high-risk location might be asked to quarantine. One category often targeted for quarantine is travellers.

In January 2020, the UK quarantined airline passengers arriving from Wuhan, China, in special facilities. The UK then abandoned quarantine for international arrivals until June, just prior to the summer holiday season, when Priti Patel – the Home Secretary – introduced a fourteen-day period of quarantine at home.

The policy was horribly convoluted because it only applied to travellers from countries deemed high risk, but with no rigorous or consistent definition of what that meant. Travel corridors with individual countries opened and closed with bewildering frequency over the summer, with the result that thousands of people changed their plans at the last minute to try to beat the latest deadline and avoid having to quarantine themselves.

Logic got lost in the confusion. The Prime Minister of Ireland announced in August that he was considering dropping quarantine requirements for arrivals from the UK (where infection rates had fallen quite low) but not from Brazil (which was suffering a major epidemic). If the aim was to minimise the total number of imported cases (variants were not a consideration at the time) then that's completely the wrong way around. Whatever the difference in the prevalence of infection between arrivals from those two countries, I am quite sure that it was dwarfed by the difference in the volume of traffic (unless there were far more arrivals into Ireland from Brazil than I ever imagined).

Introducing a requirement to quarantine is not a ban on travel, but it surely discourages it. There was, however, a less disruptive alternative to quarantine: testing, both before travelling and on arrival. I touched on this in Chapter 9 in the context of get-back-to-work tests but now we need to look at the relationship between quarantine and testing in more detail. Neither is perfect. One of the main arguments against quarantine is that people might not fully comply with the requirement to self-isolate.

The main argument against testing as an alternative to quarantine is that the tests are not 100% sensitive, particularly during the early stages of infection when levels of virus in the upper respiratory tract are still low. There were several studies estimating the risk that testing would miss an infection. The consensus was that a single RT-PCR test on arrival should catch 40–50% of pre-symptomatic or asymptomatic infections. A second test four or five days later would increase this to 80–90%. This sounded promising.

Unfortunately, a rogue analysis was published by Public Health England estimating that a single test on arrival would pick up just 7% of infections. As soon as I heard it, I knew this result was implausible. It eventually turned out to be due to some peculiar assumptions made by the analysts; they were working on the basis that the problem we were faced with was people becoming infected during their flight, a small subset indeed.

The misleading 7% figure was quoted by both Matt Hancock and Boris Johnson as approved by SAGE and the damage was done: the option of testing to reduce or obviate the need to quarantine was taken off the table. Germany, Iceland and other countries took a different path and testing of international arrivals was soon available there.

The whole quarantine versus testing debate raised an important question: were we trying to discourage travel or were we trying to make travel safe in a world where we are living with the virus? My sense is that some scientific advisors saw quarantine as a way to reduce the amount of travel as part of their objective of suppressing the virus in every way possible. Those who, like myself, were seeking ways of living with the virus saw testing on and after arrival as helping us to return to something a bit more like normality.

The challenge of relaxing restrictions

By this time, the UK administrations had arrived at a policy of gradually lifting restrictions over the summer of 2020. This appeared to be fairly successful at first. In most parts of the UK numbers of cases remained low, though there were a few larger outbreaks such as the ones in Leicester in June, Manchester in July and Aberdeen in August. This led to the imposition of local restrictions to try to suppress the virus – a targeted response that Boris Johnson referred to as 'whack-a-mole'.

Gradual relaxation was less convincing as a long-term strategy. As I explained earlier, as more restrictions were lifted then – unless we managed to suppress the virus by means other than social distancing – it was inevitable that the R number would eventually rise back above one. When this happened, localised outbreaks would become harder to contain and would eventually coalesce into a general rise in cases across the UK. (In an attempt to explain this in non-mathematical terms, my team made a cartoon animation of a whack-a-mole game where the R-above-one mole grew too big to whack).

Policy-makers were reluctant to accept this analysis, however, taking the view that if we relaxed slowly and cautiously enough then all would be well. As we would find out in September, this was a forlorn hope that betrayed a lack of understanding of the underlying epidemiology.

The sequencing of the relaxations during this period often felt arbitrary, given that the policy objective was still to keep the R number low. There were no reliable estimates of how much transmission was occurring in places like gyms, hairdressers or churches. Anyway, the models didn't have the level of granularity needed to explore the impact of opening them up or keeping them closed, so we couldn't rank them in terms of contributions to the R number (beyond broad categories such as outdoor activities are far safer than indoor). It was more a case of knowing we couldn't resume all activities, so somebody had to pick which to allow and which to keep restricted. In this respect, the scientists couldn't help much (though some were all too willing to give their views on which activities they considered more worthwhile than others, value judgements that should have been left to the politicians).

There was another way of sequencing the relaxations, based not just on the R number but also on the more traditional objective of keeping the public health burden as low as possible.

Taking this approach, the highest priority would be to open schools, given that novel coronavirus was a minimal threat to children. Next would be to open universities; again, the student population were at much lower risk. Happily, this fitted with policy-makers' desire to prioritise re-opening education.

In principle, we could have continued up the 'ladder of risk'. Contacts in managed environments frequented by younger adults where Covid-safe measures could be implemented – including many workplaces and public spaces – would be next. The losers would be activities and people nearer the top of the ladder. These include indoor socialising in people's homes and activities involving the most vulnerable. Opening up care homes would come right at the end.

A report by the economist Nicholas Stern and colleagues at the Royal Society of London looked at the lifting of restrictions through an economic lens. The report proposed 'targeted' re-opening in a way that optimises a combination of lives saved through controlling the epidemic and livelihoods preserved by increasing economic activity (though they left the tricky but central decision of how to weigh lives versus livelihoods to policy-makers). This strategy prioritised relaxations according to their contribution to preserving livelihoods. Some prioritisations were of this kind; for example, the eat-out-to-help-out scheme to support pubs and restaurants.

The relaxation of restrictions around the UK ended up with elements of all these approaches. The result was untidy and often contentious but there was no escaping the conclusion that Mike Parker – expert in ethics – was absolutely right: there was no route out of lockdown that didn't have different impacts on different segments of the population. We couldn't re-open completely so there were inevitably going to be losers as well as winners, whether they be the young, the old, care home residents, clubbers, pub landlords, airline passengers or simply the unlucky residents of the worst-affected areas, a kind of coronavirus post-code lottery.

Lockdown had hugely different impacts on different people, both when we went in and as we came out. Some people were made to shoulder a far greater burden than others. Mostly, they were not the people at greatest risk from the virus.

A longer-term perspective

The lifting of restrictions was often referred to as the route – or road map – out of lockdown. When planning a route it always helps to know where you're going. Since we couldn't eradicate the virus our destination would have to be a state called endemicity.

Endemicity means that a virus – or any other cause of infection – persists in a population indefinitely. The public health expert Peter Piot refers to endemicity as being 'part of the human condition'. When I say that we will have to learn to live with novel coronavirus I am saying that I expect it to become endemic.

Endemic viruses are kept in check by herd immunity. If we let a simple computer model run for long enough we eventually arrive at a level of infection that reflects a balance between the maximum R number of the virus (R0) and the level of herd immunity in the population. The balance exists because immunity doesn't just build up in the presence of infection, it also declines over time, perhaps because it doesn't last a lifetime or inevitably because immune people eventually die of other causes.

In real life this 'balance' can be messy, with a fluctuating prevalence of infection and perhaps significant outbreaks, though these will not be as explosive as the initial epidemic. Yet the overall trend is flat, the R number hovers around one, and the level of immunity in the population sits close to the herd immunity threshold.

It will take many years – more likely several decades – before we arrive at a fully endemic state for novel coronavirus. As it was, by the summer of 2020 we had barely begun the journey. In most parts of the world, the UK included, less than 10% of the population had been infected with novel coronavirus, so we were still far below the herd immunity threshold.

The important point here is that infectious disease epidemiologists had worked all this out as early as February 2020. This was why I was so insistent that measures we took to reduce the impact of novel coronavirus had to be sustainable. There was never any prospect of the pandemic being over in weeks or months; novel coronavirus was going to be a problem for years or decades, even with a vaccine. For most of 2020, however, many people seemed unaware of, or unaccepting of, this uncomfortable truth. That was to change in the autumn.

CHAPTER 11

SLOW-MOTION REPLAY

I knew a second wave was on the cards before we'd had the first one. In March 2020 my team at the University of Edinburgh had modelled a lockdown that ended in June and was followed by a slow, initially imperceptible, rise in cases over the summer, culminating in a second wave in late September.

We highlighted this scenario in a briefing for the SAGE sub-committee SPI-M on March 11th, twelve days before the first lockdown. We didn't present it as a prediction – a lot could happen in the next six months. We even wrote 'Not a Prediction' in large letters on our graphs. Nevertheless, we were clear that a second wave was a possibility that policy-makers should be prepared for.

In May, the Science Media Centre organised a press briefing where the epidemiologist John Edmunds and I spoke about the prospect of a second wave. In an unsuccessful attempt to not cause undue alarm I referred to it not as a wave but as a bump. I was trying to get across that though a second wave was almost inevitable it would not be as explosive as the first and therefore should be less damaging. I was right about it being less explosive, but it turned out to be more damaging than it ought to have been had we handled it better.

The epidemic in September 2020

Many others were concerned too. In July 2020 the Academy of Medical Sciences published a report that suggested that the R number could rise to 1.7 from September and the resulting second wave would cause more than one hundred thousand deaths, with a massive impact on the NHS that would be compounded by seasonal flu. This didn't sound like a plausible scenario – allowing R to rise as high as 1.7 implied the wholesale lifting of restrictions. That was possible – as I've already said, one of the hardest things to anticipate during the pandemic was what the government was going to do – but it seemed unlikely.

Even the mention of flu in the Academy's report seemed unwise. Flu has a maximum R number of no more than 1.5 – half that of novel coronavirus – and spreads in similar ways. The measures we had in place to reduce the spread of novel coronavirus should easily prevent a major flu epidemic in the coming winter. In support of this argument, Australia – whose flu season is in our summer – was seeing exceptionally low numbers of cases. I wasn't greatly concerned about flu. That expectation turned out to be correct: there was little flu in the UK in the winter of 2020–21.

SPI-M's weekly estimate of the R number did rise above one eventually – as we all knew it would given the gradual easing of restrictions – though it took a lot longer to do so than my more pessimistic colleagues had expected. My hope was that these advance warnings of a second wave would allow time for preparation and, just as important, to plan an immediate response that avoided any threat of a second lockdown. That's not what happened.

In the first week of September the UK reported over fourteen thousand novel coronavirus cases and fifty-three deaths and both were increasing. The R number was above one and climbing. Worryingly, this applied across the whole country, so it looked as though we had moved from a series of local-ised outbreaks to widespread community transmission. The epidemic was growing fastest in young adults but more slowly in the oldest age groups. If that changed then the mortality rate would accelerate too.

In mid-September the universities re-opened and quickly became the centre of an epidemiological storm. It was widely anticipated that there would be a surge in cases as around a million students relocated to communal accommodation. A team at the University of Bristol had already identified halls of residence and face-to-face classes as hotspots for transmis-sion. When outbreaks duly erupted in universities around the country there were chaotic scenes as students were quarantined en masse having only just arrived.

There are lessons to be learned from this episode, but the epidemiological damage was not as great as many feared.

First, students are almost all at low risk from novel corona-virus and the vast majority of cases were mild or asymptomatic.

Second, the universities re-opening did not start the second wave; it was already well under way before the academic year began.

Third, university outbreaks were largely contained with little evidence of spread into the wider community.

The key question is not whether re-opening universities contributed to the second wave – which it did – but whether it made a decisive contribution – which it didn't. The second wave would have happened even if the students had all stayed at home.

Universities aside, the second wave was – as expected – far less explosive than the first. Again, the key statistics are the R number and the doubling time. In mid-March 2020 the R number was around three and cases and deaths were doubling every three or four days. By mid-September the R number was around 1.4 and cases and deaths were doubling every ten or eleven days. There's a massive difference: if we started at the same level and let the two epidemics run for a month the March one would be fifty times bigger.

The implication of a slower growing epidemic was that if we implemented measures immediately we could – unlike in March – afford the time to wait to see if they worked. If we failed to bring the epidemic fully under control then we could bring in additional measures. On the other hand, we couldn't do nothing: if we let cases, hospitalisations and deaths grow exponentially we would eventually be forced to impose much tougher restrictions – and for longer – to bring them down again.

One target for restrictions in September 2020 was the hospitality sector. As with the easing of restrictions I described in Chapter 10, decisions to impose restrictions there were taken more on the basis of the judgement of public health officials than any hard epidemiological evidence available at the time. The argument was that transmission was certainly occurring somewhere, we had to do something, and the evidence was at least as good for the hospitality sector as anywhere else. As one colleague memorably put it: it's pubs or schools.

The hospitality sector was, understandably enough, far from satisfied with this rationale for picking on their industry, particularly as they'd invested heavily in making their premises as safe as possible. If we'd recognised that further restrictions were not the only option and phrased the choice as 'it's pubs or more cases self-isolating' or 'it's pubs or improved case detection' then perhaps the outcome would have been different, but that didn't happen.

A number of other measures were also introduced. The Rule of Six was intended to reduce social contacts (though it unnecessarily applied to children, who play a minor role in transmission, and outdoor activities, which are very low risk). Tighter restrictions were imposed in hotspots such as Bolton and Birmingham.

SAGE first called for an 'early and comprehensive response' to the second wave as early as September 10th. It was apparent from many media interviews that they meant a form of lockdown. I agreed that an immediate response was warranted but I disagreed that we needed to contemplate another lockdown at that stage.

On September 19th I wrote in the *Sunday Telegraph* – they made it the leading article on their web page – that our response needed to be proportionate and evidence-based, that we must have an exit strategy and that our priority should be to protect the vulnerable. This was the lesson we ought to have learned from the first lockdown: intervene early and you don't have to intervene as hard.

One option being considered in September was a circuit breaker. A circuit breaker is a form of lockdown, but with distinctive features that make it a little more palatable. A circuit breaker has a pre-determined start date and, crucially, end date. That makes it possible to plan ahead, minimising disruption to businesses and services. The idea is to deliver a short, sharp shock, reducing the incidence of new infections to a low level for a short period – two or three weeks was suggested. This would eventually feed through into fewer hospitalisations and deaths, though this would not happen until after the circuit breaker ended.

The limitation with a circuit breaker is that – as with any form of lockdown – it doesn't solve the problem, it simply defers it. My team and others in SPI-M had modelled this on-off lockdown approach back in March, so we had a good idea of what to expect. If, once the circuit breaker is over, we return to the same level of activity as before then after a while – likely to be another two or three weeks – we are back where we started. At that point we'd need another circuit breaker. In the event, that's exactly what happened in Wales when they adopted the circuit-breaker strategy.

England took a different route and devised what was known as a tier system. The idea was to impose stricter restrictions at a regional level if the R number, incidence of new cases and pressure on the local NHS rose too high. The tier scheme wasn't well received by many advisors. Chris Whitty – CMO England – was concerned that not even the top tier – Tier 3 – would be sufficient to keep R below one and the scheme would fail. A second lockdown was imposed in England before there'd been time to find out.

When the tier system was introduced on October 14th – more than a month after SAGE sounded the alarm – there were over one hundred thousand cases and over five hundred deaths per week in the UK. The calls for action had grown much louder and the clamour continued to be led by SAGE. We were subjected to a daily barrage of advisors calling for some form of lockdown and warning of dire consequences if this wasn't done immediately.

Misleading graphs and a loss of trust

The calls for a second lockdown were supported by modelling what might happen if nothing was done. You will recall from Chapter 3 that I don't think presenting the do-nothing scenario is a good approach. Just as in the first wave, there was no prospect of our doing nothing and, as I've just described, a range of measures had recently been put in place that hadn't yet been given time to work. In March 2020, there had been a lot of scepticism about the 'prediction' of half a million deaths if we didn't go into lockdown. Surely we weren't about to make the same mistake again.

Things started to go wrong on September 21st at a press briefing given by Patrick Vallance, the UK's Chief Scientific Advisor. During the briefing the CSA presented a graph showing what would happen if you started with the number of cases reported that day and doubled it every seven days. The result would be fifty thousand cases per day by mid-October. The graph was presented as an 'illustration' but, inevitably, was interpreted as a prediction. I'll call it a 'projection' because it was a simple extrapolation of the data rather than being based on an epidemiological model.

There was a lot wrong with the projection. Let's start with the seven-day doubling time. In September the epidemic was doubling every ten or eleven days not every seven days. In the event, it continued to do so and – had that figure been used – the projection would have been reasonably accurate. There was no reason to expect the epidemic to accelerate suddenly. At the very least, other doubling times could have been compared: seven days as a worst case, ten or eleven days as the current rate, and perhaps fourteen days if the measures in place had an immediate effect. It is never good practice to show a single line as if that's the only scenario possible, a lesson that should have been learned back in March.

Another difficulty with an exponential projection is deciding when to stop. If this projection had been extended for another week we would be talking about one hundred thousand cases per day. Another month would have given us close to half a million. Per day! An exponential projection will give you any number you like if you run it for long enough.

I had two concerns about the impact of this projection. On the one hand, it could precipitate an over-reaction. On the other hand – because it was so implausible – it could lead to a loss of credibility, with the result that nothing was done at all. Judging by the press reaction, the loss of credibility won out and it is entirely possible that the graph delayed action rather than hastened it.

When I first saw the CSA's graph, I quickly posted what was intended to be a reassuring comment through the Science Media Centre saying that it was highly unlikely that the UK would see so many reported cases per day by mid-October. As it turned out, we barely reached half that.

My objections did not go down well. After a flurry of e-mails I was invited to 'correct' my comments. The invitation was passed on to me by a messenger so I cannot be sure precisely where in the system it originated.

The matter didn't end there. A couple of weeks later I was asked to give evidence to a House of Commons Select Committee. This generated another flurry of e-mails over an October weekend from two senior government scientists concerned that I might criticise the CSA's graph before the MPs. As it turned out, the projection was queried by the Committee, but the question they asked about it wasn't directed at me, and was deftly turned aside.

All of this seemed rather pointless. By that time, both the BBC and *The Telegraph* had published graphics showing that the CSA's projection had not been borne out in reality. The damage was done.

The media shouldn't be given too much credit, however. Cases and deaths continued to rise through September and October but, although the underlying trend was clear, the daily data were noisy – a word statisticians use when the numbers bounce around a lot. You could always tell whether today's numbers were high or low by the prominence they got in the media. Peaks were headlines; troughs went straight to the middle pages. Unless you were paying close attention you could easily get the impression that the epidemic was growing much faster than it was.

Almost unbelievably, the September 21st saga was played out again at the October 31st press briefing by Patrick Vallance and Chris Whitty that presaged the Prime Minister's announcement of a second lockdown in England. This time the briefing included a projection by one SPI-M modelling team of up to four thousand deaths per day, several times higher than the previous peak in April. It quickly became apparent that this projection was already inaccurate on the day it was shown. Nor did it take into account the fact that the second wave was already beginning to slow.

One of the strengths of SPI-M is that it reports the results from several different models. The model that generated the four thousand deaths a day figure was an outlier – all the other model projections gave much lower numbers. The CSA was later asked by a Select Committee what it was about this model that made it so different. He didn't know. I sympathised. I'd asked the same question myself, and hadn't received a clear answer.

Still, it was correct procedure to include the outlier model in the evidence base presented to government; better that than change SPI-M's way of working at a critical juncture because we didn't like the result. That said, the episode does reinforce – if reinforcement were needed – just how important it is that the caveats, assumptions and uncertainties in models are communicated when presenting the outputs. Even more important: never rely on a single model.

As it was, the projection of four thousand deaths per day unleashed a storm of criticism, a rebuke by the Office for Statistics Regulation and a hastily called Select Committee session where the CSA and CMO were questioned about the role of models in the decision to enter lockdown.

A few days later two of the projections used in the original briefing were corrected. This time the damage was even worse.

Many people now believed that England had gone into lockdown on the basis of a 'dodgy dossier' – a term used for a somewhat similar episode in the build-up to the Iraq War in 2003.

Second lockdown

The second lockdown wasn't enforced until the first week of November. By then, there were over twenty thousand cases per day and there had been over two thousand deaths the preceding week. Hospitals in some – though by no means all – regions were treating more Covid-19 patients than they had at the peak of the first wave.

Why did the government wait so long when the alarm had been sounded in mid-September? It seemed to me that the lessons of the first wave had not been learned and that SAGE bore some of the responsibility. They were right to call for immediate action in mid-September, but that action did not have to be lockdown. Lockdown should always be a last resort – by October 2020 even the World Health Organization was saying so.

The government should have been offered alternatives to lockdown, while keeping that option on the table in case it were needed. Alternatives to lockdown were barely mentioned, something I found profoundly disappointing as there were three things we could have done in September.

First, we could have put much more effort into ensuring that novel coronavirus cases and their contacts self-isolated when they needed to. Inexplicably, we hadn't paid proper attention to the problem of low adherence to self-isolation when it first surfaced so we didn't know how to fix it. The introduction of £500 payments for those who needed financial support to self-isolate was a step in the right direction but we could have done much more. New York City set a good example. Their

'Take Care' support package for people self-isolating was so comprehensive it even included a dog-walking service.

Second, we knew from the prevalence surveys that only around half of novel coronavirus cases were being reported. This compromised attempts to reduce transmission by asking cases and contacts to self-isolate. Test, trace and isolate cannot work properly unless the cases are being found in the first place. The mass testing pilot in Liverpool in November was the first systematic – though localised – attempt to solve that problem. All in all, mass testing wasn't ramped up fast enough to have much impact on the course of the epidemic in 2020.

Finally, we could and should have done far more to protect the most vulnerable during the second wave. Chris Robertson and I sent a briefing note to SAGE revisiting our analysis showing that even if a lockdown was imposed, most people who would die would do so from infections acquired after the restrictions came into force. Lockdown may or may not be necessary, but it certainly wasn't going to be sufficient to save all the lives that could be saved. The paper never made it onto the agenda. When the UK government finally made an announcement on shielding in early November it felt, yet again, like an afterthought when it should have been at the forefront of the response.

By mid-November Scotland was alone among the UK nations in not having imposed a form of national lockdown, though some areas were living under tight restrictions. A week or so earlier, Scotland had introduced its own version of the tier system and – unlike in England – it was given time to work. The difference wasn't in the epidemiology, the second wave in Scotland developed much as it did in England. The most striking contrast to England was that there was no clamour amongst scientific advisors or commentators for a second lockdown in Scotland.

The case for the second lockdown in England remains weak to this day. The UK epidemic started to slow down in mid-October. Cases and hospitalisations were levelling off and deaths were not increasing as fast as they had been. If the epidemic were growing exponentially then not only would there be more cases every day but the number of cases would increase by more every day – that's what exponential growth looks like. This wasn't happening, indicating a slowing in the growth rate. Estimates of the R number fell closer to one nationwide, and below one in many parts of the UK. Early indicators such as the ZOE study of symptoms suggested this trend would continue. None of this could be attributed to the impact of a lockdown that hadn't even begun.

The measures taken in September and October may or may not have been enough to bring the R number below one. That's not such a problem while incidence remains low. We could have implemented measures well short of full lockdown sooner, given them time to work, and strengthened them if necessary. There was a course of action occupying the middle ground between SAGE's calls for lockdown and the government's reluctance to implement that option. We didn't go down that route and ended up with more cases, hospitalisations and deaths, and lockdown: the worst outcome of all.

CHAPTER 12

NEW VARIANT

I have already referred to the publication of the complete genetic code – or genome – of the novel coronavirus in January 2020 as a pivotal moment. It allowed the development of RT-PCR diagnostic tests and precipitated initial work on vaccines. It was also the first step towards being able to track the virus as it moved and as it evolved, a capability that was to become a pivotal part of the UK's pandemic response.

Genome sequencing

Novel coronavirus's genetic code is written using four 'letters': G, C, A and U that correspond to different molecular subunits called nucleotides. The nucleotides are arranged in a sequence to make up a strand of RNA, which is the genetic material of some viruses. (The more familiar DNA molecule – used by other viruses and all other living organisms – is a sequence of a slightly different set of letters: G, C, A and T). The novel coronavirus genome is about thirty thousand nucleotides long and contains just a dozen genes. The technique we use to read the genetic code of the virus – or anything else – is called genome sequencing.

Genome sequencing has been around since the 1990s but was brought to the world's attention when the first human genome was sequenced in 2003. That task took several years and a massive effort involving a consortium of major laboratories around the world. The technology has moved on since then – we now have portable sequencing devices the size of a mobile phone that can sequence a virus genome in hours and can be used almost anywhere. The use of these devices is routine for my research team but I am still mesmerised by the power of this technology.

Virus genomes are relatively small and we can sequence large numbers of them quite easily. Thanks to the accumulation of mutations, every genome is different but more closely related genomes are more similar. This is the basis of phylogenetics, a method for drawing up a family tree. We can use a virus's family tree to track its spread in close to real time. We can re-trace the virus's path from country to country, city to city and even – in some instances – from person to person.

The UK has by far the biggest novel coronavirus genome sequencing programme in the world. The Covid-19 genomics UK consortium – COG-UK – was set up in March 2020 under the leadership of Sharon Peacock at Public Health England. Sharon had spent years pioneering the use of genome sequencing as a public health tool – COG-UK was the culmination of her efforts.

There are strengths and weaknesses to the approach. To put it simply: genomics can give us unique insights into what has happened, is less useful for telling us what is happening right now and cannot tell us what is going to happen next. In other words, it gives a retrospective not prospective view on the course of an epidemic.

Two practical issues that have limited the usefulness of genomics in the past are speed and scale. COG-UK pushed both to the limit. It didn't quite operate in real time, but it was fast: a two-week lag from sample to sequence was the norm in the first year of the pandemic. It also operated at an impressive scale: by the end of 2020, COG-UK had sequenced well over one hundred thousand viruses and had contributed more than half of all the sequences deposited in the global databases.

Learning from genomics

Early reports from COG-UK told us that novel coronavirus had been imported into the UK at least a thousand times in the first quarter of 2020, the majority from Spain and France. Only a tiny fraction came from China. However, the number of imports turned out to be less important than the date of their arrival. As we might expect (because we know that exponential growth beats linear growth), the earliest arrivals seeded the largest number of cases. The arrival of new viruses slowed during lockdown but by then, of course, the damage was done.

Another COG-UK study looked at novel coronavirus genome sequences from Scotland over the first eight months of the pandemic. One finding was that few viruses had been brought into Scotland from England over the summer. Those who had wanted to ban tourists had been misguided – the majority of cases in Scotland that summer were caused by virus lineages (a lineage means a direct line of descent) that had been present since March.

However, around a quarter were viruses from mainland Europe. The most likely source of these viruses was not visitors, but Scottish residents going on overseas holidays. Epidemiological studies showed that overseas travel was a risk factor for infection during the summer (and it would be no surprise to learn that this was at least partly due to tourists taking a 'holiday' from social distancing too). There had been less talk about banning Scottish residents going on holiday than there had been about banning visitors from England, but it would have had the greater public health impact.

The COG-UK study also looked at the viruses that were spreading during the autumn second wave in Scotland. There had been claims that the second wave was driven by infections among arriving university students. As we saw in Chapter 11, this idea was not supported by epidemiological studies. Though there were several large outbreaks in Scottish universities, these remained largely confined to campus and there was little evidence of wider spread. The genome sequence data confirmed this: fewer than 1% of cases were attributable to virus lineages recently brought into Scotland. The university students had been caught up in Scotland's second wave, but they did not cause it.

There was, however, a twist to the story the genomics told. The second wave in Scotland was mainly made up of the descendants of those viruses that had been imported during the summer, most likely by people returning from holiday. The commonest lineage – called 20A.EU1 – came originally from Spain. We would call it a 'variant' now, but the term wasn't in common usage at the time. It is possible that 20A.EU1 viruses were slightly more transmissible, but they may just have been in the right place at the right time – what genome scientists call a founder effect and the rest of us call luck.

Far more importance was attributed to this finding than it warranted due to an unfortunate misunderstanding. The COG-UK report claimed that novel coronavirus had almost died out in Scotland during the summer of 2020. What the authors meant was that they didn't have many virus genome sequences from this period, which isn't the same thing at all. This was hardly surprising given that we were finding only around 10% of cases during that phase of the epidemic, and genome sequences from only a fraction of that 10%.

When this large gap in the data is taken into account, it looks as though most or all of the commonest virus lineages in

Scotland persisted from the first wave to the second. Whether people had been allowed to go on overseas holidays or not, there would still have been a second wave in Scotland, as there was in every western European country regardless of their policy on international travel.

The myth that Scotland came close to eliminating novel coronavirus in the summer of 2020 lived on regardless, even though – as we saw in Chapter 10 – the epidemiological evidence didn't support it either. I suspect this was partly due to an eagerness to see novel coronavirus as an external threat rather than a home-grown problem. As a result, the contribution of long-distance travel to the second epidemic wave – beautifully captured by the phylogenetics studies – got exaggerated and the contributions of, say, the re-opening of restaurants or the resumption of family gatherings were downplayed, even though they will have been far more important.

The lesson here is that we have to consider all kinds of evidence and not just evidence from one source, however impressive the technology.

A not-so-merry Christmas

The second wave of novel coronavirus in England culminated in the November 2020 lockdown. Almost as soon as the new lockdown began questions were being asked about what would happen when it was due to end and, in particular, what could we expect for the Christmas holidays.

By now, it was apparent to everyone that the latest lockdown – just like the previous one in March – would not solve the problem, it would just defer it. Several epidemiologists suggested that it would not even do that and by the Christmas holidays the situation would not be much better than it was when the lockdown began.

Whether the lockdown succeeded in lowering incidence

or not, there was a lot of concern that anything like a normal Christmas – combining travel with parties and social visits, including multi-generational family get-togethers – could generate a surge in Covid-19 cases. A now familiar parade of scientists appeared in the media calling for Christmas to be cancelled altogether. There were even doubts as to whether university students would be allowed to go home at the end of term.

The concern was legitimate but the evidence didn't fully back it up. The same arguments had been made about the Thanksgiving Day holiday in the US. The raw data were hard to interpret: there was a spike in cases – but not deaths – after Thanksgiving, but any upward trajectory began well before the holiday and the pattern was inconsistent between states. The BBC's Reality Check team concluded that there was no Thanksgiving Day effect. Some US epidemiologists disagreed but no formal analysis was published in a scientific journal.

Another argument for a Christmas surge was the well-established pattern of a spike in influenza cases around the New Year period. This happens, but not consistently, and at the same time there is also a spike in most other conditions that send people to hospital, not just infectious diseases. This pattern can be explained by people not seeking, or not being able, to access health care until after the holidays. Most general practices see a smaller-scale version of the same effect in their waiting rooms every Monday morning.

Evidence from behavioural studies shows that contact rates generally fall over the Christmas holidays rather than rise. There is – as you'd expect – an increase in social mixing, but this is more than outweighed by people not going to work or school. Despite this, there was a specific concern about increased social mixing between young people – who at that stage of the epidemic had the highest prevalence of infection – and the elderly. Discouraging inter-generational mixing could have a public health benefit, though it would come at a significant social cost.

As always, there was another way: we could focus not on reducing the number of contacts but on making those contacts safe. This idea did not gain hold in what became an increasingly hysterical debate.

The November 2020 lockdown itself turned out to be more successful than many had expected, particularly those who still believed it was impossible to control the epidemic while schools remained open. The incidence of cases had fallen by almost half as we entered the final week of lockdown. Then there was an unexpected change in the epidemic's trajectory. Nationally, incidence levelled off, even though the lockdown was still in place and there had been no marked change in patterns of behaviour. More surprising still, in London and south-east England incidence started to climb.

Several unconvincing theories – some revolving around Christmas shopping – were put forward to explain this unexpected increase. The true explanation was something of a surprise.

Variant of concern

The first I heard of a new variant of novel coronavirus was when Matt Hancock announced its existence in parliament on December 14th. Genome sequencing had identified a variant designated VOC 202012/01 (VOC stands for variant of concern). He claimed that this new variant could be associated with faster rates of spread in the south-east of England. Public Health England were investigating and the World Health Organization had been notified. All of this was news to me.

I immediately contacted colleagues who worked with COG-UK. They were as surprised as I was. They had known about the variant for some time but had not thought it particularly significant and did not know why it was suddenly being highlighted.

I did learn that reports of the new variant were emanating specifically from the Lighthouse laboratories. The Lighthouse labs were a network of seven high-throughput novel coronavirus diagnostics facilities set up during the pandemic and they worked a little differently from other testing laboratories. That difference turned out to be important.

A feature of the new variant was that a tiny section of the genetic code for the spike protein had been deleted. This happened to stop the RT-PCR test from working, but just for that gene – the S gene. Only the Lighthouse labs tested for the S gene; other diagnostic labs did not. The Lighthouse labs also tested for two other genes which meant they could identify the new strain as an 'S gene dropout'. The new variant wasn't the first S gene dropout but it had become by far the most common. The S gene dropout characteristic was useful because it made this particular variant much easier to track – we didn't need to sequence the genome of every case or develop and deploy a tailor-made diagnostic test.

That there was a new variant didn't necessarily mean it was the reason for the recent increase in cases. I didn't get to see the evidence supporting that claim until I was invited to a special meeting of NERVTAG – another advisory group feeding into SAGE – on December 21st.

As soon as I read the papers the day before the meeting I became much more concerned. It looked as though the new variant was spreading roughly 50% faster. That's a huge increase and enough to raise the R number above one even during a lockdown, consistent with the puzzling upturn in cases earlier in the month. The new variant was most common in the same geographical areas that were seeing the fastest increase in cases: London and east and south-east England. The most obvious explanation was that the new variant was indeed more transmissible.

Given that the new variant had already been detected in Glasgow, I highlighted these findings to my colleagues in Scottish government over the weekend ahead of the NERVTAG meeting and advised that immediate action was called for – there was every prospect of an explosive increase in cases over the next week or two. There was: cases in Scotland had tripled by the end of December, hitting an unwanted record of well over two thousand cases per day.

It was a busy Christmas period as meeting after meeting was called to discuss the new variant and its implications. It took a while to convince my colleagues in Scotland that we should take the new variant seriously. Their reaction was just the same as mine had been: there are many different variants of novel coronavirus, so we need compelling evidence to consider this one as a variant of concern. The conundrum was that if we did need to act then we needed to act quickly. 'Waiting to be sure' was a big risk; cases due to the new variant were doubling every seven days. The only time we had seen faster growth was back in mid-March before any restrictions were put in place.

Changing phenotypes

Of itself, the appearance of a new variant is no surprise. Novel coronavirus started to diversify soon after it first emerged and – depending exactly how you define 'variant' – by the end of 2020 there were thousands all around the world.

One of these was 20A.EU1, which we encountered earlier in this chapter. 20A.EU1 was a variant that was first detected in Spain in June 2020 and arrived in the UK later that summer. Other variants followed. To help us keep track, various naming systems were developed: VOC 202012/01 – which arose within the UK – was re-named once as B.1.1.7 and once again as the alpha variant, which is easier to remember, though many people still refer to it as the UK or Kent variant.

Usually, the emergence of a new variant makes no difference to the way the virus behaves and so has no public health implications. The existence of variants helps us to track the spread of the virus and that's about their only significance. It's the rare variants that do behave differently that are a concern. Geneticists call this a change in phenotype, meaning measurable characteristics of the virus. There are three characteristics that we are particularly concerned about: transmissibility, pathogenicity and immune escape.

Transmissibility – as we've seen – is about how readily the virus passes from one person to another. The alpha variant has mutations that alter the spike protein in ways that increase its affinity for the ACE-2 receptor. This allows higher levels of virus to build up in the upper respiratory tract, making an infected person more infectious to others. That doesn't mean the virus spreads in a different way, so the same interventions to reduce the risk of infection still work, but they need to be implemented with even greater stringency.

The alpha variant was 50% more transmissible, which led to it becoming dominant across much of the world within a few months. Many – including me – had been surprised by this and well into 2021 prominent virologists were disputing that such a large increase in transmissibility was plausible and arguing that there must be some other explanation for the data. None came to light.

Pathogenicity means the likelihood that an infection will cause severe illness, which we'd expect to translate into a higher infection fatality rate as well. There was a lot of debate about whether or not the alpha variant was more pathogenic as well as more transmissible. Early reports suggested it was and two papers published in March 2021 estimated a mortality rate over 60% higher than older variants. Public Health England then published their own report suggesting that alpha wasn't more deadly. This result was hotly disputed.

So it seemed that the alpha variant was more transmissible and was probably more pathogenic too, a worrying combination. This unwelcome turn of events did at least put to rest the idea that as novel coronavirus evolved it would become more benign. This proposition comes from a decades-old, but poorly supported, theory that a virus – or any other cause of infectious disease – will be more successful if it doesn't harm its host. I was never convinced by this line of reasoning myself. Sick people transmit novel coronavirus perfectly well, particularly to those caring for them. If they die, they usually die after the infectious period is past, so there is no downside for the virus. I was never optimistic that evolution would work in our favour.

Immune-escape variants are perhaps the most worrying of all. Immune escape occurs when someone previously infected is re-exposed and their immune response fails to recognise – or only partially recognises – the new variant. This would mean that people could get re-infected, that a vaccine might not work and – if immune-escape variants arise frequently – that it would become harder or impossible to maintain high levels of herd immunity.

Two early partial immune-escape variants were beta and gamma. The beta variant was linked to a second wave of infections throughout southern Africa and to disappointing results of some vaccine trials in South Africa. Gamma was linked to a major resurgence of infection in the already hard-hit city of Manaus in Brazil. Both variants have mutations in the spike protein, which is one of the main targets for the immune response.

As more and more variants were discovered, an interesting

pattern emerged. Some mutations turned up repeatedly and independently in completely separate novel coronaviruses lineages. Two prominent examples are mutations in the spike protein gene called N501Y and E484K (the numbers refer to where the mutation is in the gene and the letters refer to the change it makes in the spike protein). Some virologists have suggested that this pattern implies there will ultimately prove to be only a limited number of successful variants. Let's hope so, but we were nowhere near that point in late 2020.

Tackling new variants

Following the discoveries of alpha, beta and gamma, the list of variants of concern quickly grew to over a dozen. The realisation that new variants were arising all over the world and spreading far and wide re-ignited the debate about closing borders. There was good reason for this. We saw in Chapter 10 that international travel restrictions make most sense if they are put in place before a virus becomes established within your own borders. That argument works for variants too.

The problem is that it takes time to confirm an association between phenotype and genome sequence, so a new variant may already be widespread by the time we know we should be worried about it. According to phylogenetic analyses, the alpha variant emerged in the UK in September 2020 and was circulating for months before it was linked to increased transmissibility. In that time it reached as many as fifty other countries. Even during a pandemic the world is so interconnected that a new variant can spread globally within weeks.

In practice, when alpha was identified as a variant of concern many countries did put restrictions on travel to and from the UK. This led to chaos in Dover over Christmas as truck drivers were prevented from crossing by ferry to France. This was resolved in a few days by the sensible expedient of testing the drivers before they boarded. The border closure at Dover was a knee-jerk reaction to a newly recognised threat and it came too late, the alpha variant was already present on mainland Europe and would soon become the dominant variant there too.

If the UK – with its huge genome sequencing programme – could not detect a new, more dangerous variant before it has spread beyond our borders then there was little chance that countries with more limited sequencing capacity could do so. Beta and gamma had both arrived in the UK before they were identified as variants of concern. Waiting until that happened before introducing travel restrictions was proving a far from fool-proof way of keeping problematic variants out.

If reactive, targeted travel restrictions don't work then the alternative was a comprehensive ban on travel from any country where the virus was present. Even that strategy was unlikely to succeed indefinitely because essential cross-border traffic – including food shipments – would need to continue. The epidemiological advice was that travel restrictions would delay but not prevent the importation of new variants – the delay could be months, weeks or mere days, depending on the exact circumstances.

Nor was there an obvious exit strategy. Novel coronavirus would continue to throw up new variants – this could happen at any time, anywhere the virus was circulating. The virus wasn't going away any time soon, so the borders would have to stay closed. Australia was already talking about not re-opening its borders until 2022, and even that looked optimistic.

I am sure that by now you've spotted the similarity with the dilemma over lockdown. Just like lockdown, closing the border is a crude intervention that doesn't solve the problem, it just defers it to another day. The best that closing our borders can do is buy us time. Unless we plan to keep our borders closed for ever, the key question is what we do with the time we have gained.

The only sustainable strategy is to find ways to make travel safe. That can be done by testing, already in place for airline passengers in 2020 and – as we saw in Dover – easily extended to other cross-border travellers. Once vaccines became available another option was vaccine passports. Some tour operators were already planning on this basis before the end of 2020.

In the meantime, the UK had to deal with the more transmissible alpha variant that was firmly established by the end of 2020. For a start, the variant gave all the UK administrations a valid excuse to cancel Christmas, which ended up being reduced to limited get-togethers on Christmas Day itself.

In the event, there was so little mixing over Christmas that one colleague commented that it was as if we had put ourselves back into lockdown. There were a few behind-the-scenes remarks that the alpha variant had appeared at a suspiciously convenient time for policy-makers contemplating a U-turn. On this occasion, I can vouch for the government; alpha was real, it was a problem and it was the Grinch that stole Christmas.

Cases continued to rise over the Christmas period at much the same rate as before. Pressure on the NHS continued to build and – thanks to the lag between infection and hospitalisation – this trend looked set to continue until well into January 2021, if not longer. I couldn't see any way this would end without a third lockdown. I'd thought that from December 20th when I first saw the data showing how fast the new variant was spreading, so fast that a November-like lockdown might not be enough to bring the epidemic under control. There would be a clamour for tighter restrictions, to close schools and to close the borders too.

Even with a more transmissible variant it should have been possible to avoid such drastic actions if we had prepared for further waves by putting the right measures in place in the second half of 2020. We needed better case finding, better support for and enforcement of self-isolation, more investment in Covid-safe measures backed up with mass testing, and – above all – better protection for the most vulnerable.

We hadn't done any of that and so, yet again, lockdown was going to be the only option on the table. The good news was that – as Matt Hancock liked to put it – the cavalry was on the way, in the form of vaccines.

CHAPTER 13

THE CAVALRY

Vaccines have been extremely successful in protecting us from viral diseases, not least by giving us the means to control and then eradicate smallpox, but also by greatly reducing the burden of infections such as measles. Such success stories, however, are of vaccines developed and deployed over decades. Aiming for the same kind of success against novel coronavirus in a matter of months was asking a lot, to put it mildly.

Nature of immunity

To understand how vaccines work, we need some more background about immunity. Let's start with the example of measles, a virus that we understand well. Once you've had measles your immune system becomes primed to respond rapidly and effectively should you ever be exposed to the measles virus again. This is acquired immunity. Acquired immunity to measles gives us close to complete protection against reinfection for the rest of our lives. Hardly anyone gets measles twice. Vaccination against measles – nowadays most commonly delivered through the MMR vaccine, which also protects against mumps and rubella – is similarly effective.

From the onset of the pandemic, scientists hoped that acquired immunity to novel coronavirus would turn out to be as effective it is to measles. A highly effective immune response means better protection for individuals and – at the same time – a bigger contribution by herd immunity to damping down the epidemic in the long term.

Unfortunately, not all infections generate measles-like immunity. Immunity to another common virus – respiratory syncytial virus – builds up over multiple infections. This partial immunity can still be helpful. If reinfection does occur then people who have previously been infected may not get as ill second, third or fourth time around. This also happens for a variety of other infectious diseases including malaria. Another issue is that not all acquired immunity lasts for life. Protection against reinfection decays over time for many bacteria and a number of viruses too.

To add to the list of potential concerns, some viruses have ways of getting around acquired immunity. Influenza is a good example. Influenza viruses are problematic because the most important antigens – the proteins on the surface of the virus that stimulate an immune response – are continually changing, through two mechanisms known as antigenic drift and shift. As a result, they don't elicit such a strong immune response in people who have already been infected with an earlier version of the virus. This is why – unlike measles – you can get flu more than once and we have to change the flu vaccine every year.

In early 2020 we had only a few small hints of the kind of immunity a novel coronavirus infection would generate. We knew that people can get infected with some of the other human coronaviruses multiple times. We also knew that the spike protein – the most immunogenic part of the virus – is highly variable. That raised the prospect of immune-escape variants even before any were found. This possibility was reinforced by a trickle of reports over the course of the year of individuals being infected with novel coronavirus a second time. There were too few to be a major concern but they were a warning that we shouldn't be over-optimistic.

The bottom line is that throughout the first year of the pandemic we did not know how effective naturally acquired immunity to novel coronavirus would turn out to be. We had no idea at all how well a vaccine might work, even if we could develop one in record time.

Vaccine development in record time

Work on novel coronavirus vaccines began as soon as the genome sequence of the virus was published by Chinese scientists in January 2020. Global pandemic preparedness initiatives such as the Coalition for Epidemic Preparedness Innovations – known as CEPI – had been set up some years before to accelerate vaccine development for novel viruses. As the CEPI website puts it: vaccine development has historically been a slow, risky and costly endeavour. That needed to change.

The one thing in favour of developing a novel coronavirus vaccine was the enormous effort invested in the task. Over the course of 2020, work progressed around the world on over a hundred vaccine candidates. Dozens got as far as clinical trials. There were several different types: recombinant, killed virus and attenuated virus vaccines. Two of the early leaders in the race – the Moderna and Pfizer vaccines – were a completely new type, called mRNA vaccines. New technology or old, biomedical science and big pharma were delivering when we most needed them.

Any candidate vaccine has to pass a series of rigorous safety and efficacy checks before it is licensed for use. The first checks in humans are called Phase I and Phase II trials. The vaccine is given to a limited number of healthy volunteers who are monitored closely for any adverse side effects. Even before this stage, the vaccine is tested in animals so it is rare for a Phase I or II trial to reveal major safety problems, though it can happen. At the same time, the immune responses generated by the vaccine are also closely monitored. By comparison with the response to a natural infection, we get an early indication of whether the vaccine is likely to be effective.

In Phase I and II trials the vaccinated subjects are not exposed to infection. There was plenty of discussion about whether so-called challenge experiments – deliberate exposure of vaccinated volunteers to infection – should be allowed, in order to speed the vaccine development process along. None were conducted in 2020, though some were approved the following year.

The next stage is a Phase III trial. Now the vaccine is given to a large number of subjects – thousands or tens of thousands – and a similar number receive a dummy injection, a placebo. Placebo and vaccine are allocated at random, with no-one knowing who has had which until the trial is complete. Then it's a question of waiting until a pre-designated number of people get symptomatic infections, at which point the trial is over and the result is revealed. If the vaccine works then there will be many more cases in subjects given the placebo than in those given the vaccine.

We need to know precisely what 'if the vaccine works' means. The novel coronavirus vaccine trials were designed to look for protection against symptomatic infection. Vaccine efficacy is then estimated as the proportion of symptomatic infections prevented. Let's suppose that there were one hundred and ten symptomatic infections by the end of the trial, ten in the vaccinated group and one hundred in the equal-sized placebo group. Given this outcome, we would reckon that the vaccine has prevented ninety out of an expected one hundred cases in the vaccinated group, and so has an efficacy of 90%.

Once the full safety and efficacy data from a Phase III trial are available, the next stage is for the regulators to decide whether or not to licence the vaccine for use. The entire process from initial development to licencing is minutely prescribed and is designed to be as rigorous and robust as possible. Given that the ultimate goal might be to vaccinate millions, even billions, of people as quickly as practicable, safety was paramount and it was imperative that novel coronavirus vaccine trials were conducted to the highest standards.

In a departure from normal practice, in November 2020 both Moderna and Pfizer reported their interim results, usually just a check to see if the trial is worth continuing and if it is safe to do so. The results were encouraging and attracted a huge amount of attention, but we would have to wait until the following month for the definitive results. The final trial data reported by Moderna and Pfizer in December confirmed that both vaccines were safe and indicated that they gave well over 90% protection against developing symptomatic infection. This was wonderful news and prompted much rejoicing in the media.

The trial results were certainly encouraging, but related only to the trial outcome, symptomatic infections. The trials didn't tell us the three things we most wanted to know.

First, we wanted to know how effective the new vaccines were at preventing severe disease and so reducing the burden on the NHS. Everyone hoped that would be over 90% too.

Second, we wanted to know how well the vaccines prevented infection and blocked transmission. That would tell us whether the new vaccines might get us over the herd immunity threshold – herd immunity is all about infections, it has nothing (directly) to do with disease.

Finally, we wanted to know how long the beneficial effect of vaccination lasted, hoping that it was at least a year and preferably much longer.

The trials weren't designed to provide the answers to any of those questions, so we would have to wait until the vaccine was rolled out at scale. Epidemiological studies carried out after a vaccine is approved and rolled out are known as Phase IV trials. As one colleague put it, we were about to embark on the largest and perhaps most important Phase IV trial in history.

Planning a route to freedom

The Moderna and Pfizer vaccines were licensed for use in the UK in early December 2020. They were joined by the AstraZeneca vaccine at the end of the month. The vaccines themselves were produced in remarkably quick time but those successes didn't come from nowhere – they were the outcome of many years' effort and investment. The AstraZeneca vaccine was developed near Oxford at the Jenner Institute, which was founded in 2005. I know Adrian Hill, the institute's director, and Sarah Gilbert, who led the novel coronavirus vaccine research. Both have devoted their careers to developing new vaccines and – thankfully for all of us – were perfectly placed to rise to the challenge during the pandemic, and did so.

The efforts of the vaccine developers were complemented by quick work from the UK's Medical and Healthcare products Regulatory Authority, abbreviated as MHRA. The MHRA was able to authorise the new vaccines for general use – a process that can take years – at unprecedented speed because they had been working alongside the vaccine developers from the earliest stages. Mass vaccination of adults began in the UK before Christmas 2020, just under a year after novel coronavirus was first recognised, an astonishing achievement.

A hotly debated question in late 2020 was who should be vaccinated first. If we wanted to bring down the R number – suppress the virus – as quickly as possible then young adults should be prioritised, since they contribute most to transmission. Alternatively, the greatest public health benefit would come from vaccinating those who are most vulnerable to novel coronavirus and contribute the bulk of the burden on the NHS.

The second strategy made most sense if it turned out that the vaccine was better at preventing disease than it was at generating herd immunity. The Joint Committee on Vaccination and Immunisation opted for this approach when they published their vaccination priority list in December. Their decision was supported by many public health experts who had dismissed the idea that we should take special measures to protect the most vulnerable in other ways. That seemed inconsistent to me, but we got there in the end.

The Joint Committee on Vaccination and Immunisation's top seven priority groups covered approximately sixteen million people over sixty-five years old or otherwise clinically vulnerable. Many commentators – and some of my fellow scientists – imagined that once these people had been vaccinated the epidemic would be effectively over and normal life could resume by Easter 2021. Sadly, it was never going to be that simple. The problem was that not all those people would be fully protected: the vaccines were unlikely to be 100% effective, and on top of that some might not be willing to be vaccinated at all.

Let's plug in some numbers. Imagine that of those sixteen million vulnerable people 95% were vaccinated – which is the World Health Organization's target coverage for mass vaccination programmes – and that the vaccine provided 80% protection against severe disease leading to hospitalisation. That looks like a pretty good performance, but it still leaves the equivalent of well over four million people at risk. To give some context, if just half that number of people got infected (corresponding to an entirely plausible attack rate of 50%) we could end up with more Covid-19 patients in hospital than we'd had in the entire pandemic so far.

The implication of this calculation was stark and disagreeable: there was a real risk that the NHS could be overwhelmed again, even with a vaccine, if we abandoned all countermeasures. The next question is: if we couldn't fully relax restrictions having vaccinated the great majority of the most vulnerable then when could we? In December 2020 there was not enough evidence to answer that question. We would still be debating it well into the following year.

Promises, promises

This is a good moment to pause and reflect on what was promised and delivered over the course of 2020. As the UK went into lockdown on March 23rd 2020 no-one knew when – or even if – a vaccine would become available. On the downside, no vaccine had ever been developed from scratch in less than four years, and many years' work on vaccines against two other coronaviruses – SARS and MERS – had not been successful. On the other hand, never before had there been as great an incentive – nor as great an effort – as there was in 2020. Even so, most experts in the field were talking about at least a year before we had a vaccine against novel coronavirus.

Thankfully, vaccines did become available within a year, but it was a huge gamble to rely on that happening. In Chapter 3, I discussed planning for the reasonable worst case – that's not controversial. Planning for the reasonable best case, on the other hand, would normally be regarded as grossly irresponsible. Yet that's what we did. Matt Hancock has argued that the gamble was justified because it paid off. I disagree: I do not believe that most people fully understood the terms of the wager.

It was misleading to imply that when a vaccine became available this would augur an immediate end to the crisis. At best, it would take many more months for a mass vaccination programme to be rolled out and for society to return to anything like normality. At worst, if the vaccines gave only modest protection against severe disease and had little impact on transmission, we'd still need countermeasures in place – perhaps including social distancing – even once the roll-out was complete.

None of this was widely understood when the UK first went into lockdown in March 2020. Throughout that year, reassurances that restrictions would only be kept in place for a limited time were bolstered by a drip-feed of optimistic bulletins from vaccine developers. By September – when Matt Hancock was describing the UK government's strategy as getting through the winter until the cavalry arrives – we were being told that the end was in sight. It wasn't. Some kind of end – or, at least, a beginning of a new phase – of the epidemic only came into sight in December when the first vaccines were licensed for use.

False expectations do more than dash hopes, they also influence policy making. As we saw earlier, the shorter the period we expect to wait for an effective vaccine the easier it is to justify strict lockdown measures. The promise of a vaccine that would quickly bring an end to the pandemic underpinned the eagerness of scientific advisors to recommend lockdown, the willingness of politicians to introduce such a damaging policy and the acquiescence of the public. The corollary is that if we have to wait longer than expected then the justification is undermined.

As it turned out, we would be living with restrictions not for the few weeks suggested in early March 2020 but for well over a year. If we had properly planned for so long a wait before the cavalry came to our aid – and recognised that the fight would not be over even then – we would surely have made different decisions in the interim and less harm would have been done.

Roll-out and coverage

Thankfully, the UK's vaccine roll-out went well. By mid-March 2021 the NHS was vaccinating up to half a million people per day and the top seven priority groups had all been offered at least one dose of either the Pfizer or AstraZeneca vaccine. By the end of April over fifteen million people had received both doses. It was expected that all adults would receive at least one dose by the end of July.

The number of people vaccinated is the headline measure of progress, but – as my calculations a little earlier show – coverage was also critical. At the end of 2020 the signs had not been encouraging. Polling data suggested significant levels of vaccine hesitancy with up to a third of the adult population saying they might not get vaccinated. This changed dramatically as the roll-out began. Among the highest priority groups fewer than one in twenty were not being vaccinated.

Maximising coverage was important because there is a big difference between a vaccination programme with, say, 96% uptake and one with 98%. That small increase in uptake halves the unvaccinated population (from 4% to 2%), so greatly reducing the number of vulnerable people at risk of serious illness. This was a strong incentive to make sure that the novel coronavirus vaccination programme extended as far as possible into the hardest-to-reach populations. We couldn't afford to leave anyone behind, for their sakes and for everyone else's too.

So far good, but we still didn't know how effective the vaccines would be at preventing severe cases. The EAVE project set up by my colleague Aziz Sheikh many months before was about to pay dividends. Having won a long battle with bureaucracy to be allowed full access to data for almost the entire population of Scotland – more than five million people – his team was now able to estimate how much less likely a vaccinated individual was to be admitted to a Scottish hospital with Covid-19 than an unvaccinated person in the same risk category. The answer was over 80% from just a single dose from either vaccine (it would be even higher from two doses). This encouraging result was published in February 2021 and quickly replicated elsewhere.

The next questions were how well the vaccines prevented infection and transmission. These were answered by a Public Health England study called SIREN. SIREN studied health care workers – who were early recipients of the vaccines – and estimated over 60% protection against infection. The same study was able to estimate the impact on transmission using data from contact tracing studies. Cases who had been vaccinated were a little over half as likely to transmit infection to members of the same household as cases who had not been vaccinated. Combining those two results we get an estimated reduction in transmission in a fully vaccinated population of around 75%.

It still wasn't clear whether 75% would be enough. If the herd immunity threshold was at the 67% mark – the figure we arrived at earlier – then it could be, as long as we vaccinated enough of the population. Unfortunately, the alpha variant had raised the bar. The vaccines were effective against this variant but its higher transmissibility meant that the herd immunity threshold was now around 75%. That this was the same figure as the reduction in transmission was complete coincidence – the two numbers are unconnected and could be quite different – but it meant that we were right on the cusp. Getting to the herd immunity threshold was going to be difficult, and perhaps just out of reach.

Fortunately, there were three other factors in play that would help.

First, approaching 20% of the population had already been infected with novel coronavirus and would have some natural immunity.

Second, if we were close to the herd immunity threshold then even limited Covid-safe measures and small changes to our behaviour could make a decisive difference.

Third, for technical reasons I won't go into here, I suspected that the 75% estimate of the reduction in transmission was too low. Putting all this together, by March 2021 it looked to me as though the herd immunity threshold was in reach. The two caveats to this optimistic conclusion were duration of protection and variants.

We simply didn't know how long vaccine-induced protection – or natural protection – against novel coronavirus would last: it could be months, years or a lifetime. The reasonable worst case was that it decayed so quickly that even an annual vaccination campaign wouldn't be enough to maintain levels of herd immunity. Some commentators were concerned by studies showing that some cases had no detectable antibodies just a few months after being infected. Immunologists were less concerned because antibody levels are expected to decay following an infection (or vaccination); that doesn't matter as long as the immune system is primed to produce more if needed.

By May 2021, there were encouraging reports that the cells responsible for long-term immunological memory were present in people who'd had natural infections, though no-one could yet say whether the same would be true following vaccination.

Variants were another concern. As we just saw, if we were close to the herd immunity threshold then even small changes in our behaviour could make a decisive difference. It would be the same – but in the wrong direction – for small changes in the virus. If a new variant was yet more transmissible that would raise the herd immunity threshold again. Another possibility was that the vaccines were less effective at protecting against infection or disease caused by a new variant.

Most virologists were sceptical that complete vaccine escape was likely, or even possible, but partial escape was. A report published in March 2021 suggested that the AstraZeneca vaccine had limited effectiveness against the beta variant. This wasn't a wholly convincing study – the numbers of cases were too small – but it caused a great deal of anxiety.

The obvious solution would be a booster vaccination campaign using vaccines designed to protect against any problematic variants. This would take several months to implement, so it would be important to decide that such a step was needed at the earliest possible stage.

Virus genome sequencing would be a vital component of any early warning system, not just in the UK but worldwide. However, tracking variants of concern would only be possible if we knew they were variants of concern in the first place. We were going to need laboratory and surveillance systems in place to establish this much faster than we had managed in 2020.

Concerns about safety

There was one other potential pitfall. Novel coronavirus was a serious risk to a substantial minority of mainly older people, but much less so to the rest of the adult population, and hardly any risk at all to healthy children. This would not be an issue if the vaccines were 100% safe, but no medical product is 100% safe – there is no such thing as zero risk. The question of balancing risks was bound to arise sooner or later.

In March 2021 small numbers of people who had received the AstraZeneca vaccine were found to have blood clots in the brain and a blood condition called thrombocytopenia. These are extremely rare conditions so any cluster of cases attracted attention. There was a lively debate about whether there was any link with the vaccine, and plenty of confusion as different European countries made different recommendations about who should and should not be given the vaccine.

By late April in the UK, thirty-two people had died as a result of blood clots. The MHRA recommended that people under thirty years old should be given an alternative vaccine to AstraZeneca where possible. They stressed that the risk was extremely low, less than one in every half a million people vaccinated. For the young and healthy, however, it was no longer a given that the risk outweighed the benefits. This was a blow to the UK's vaccination programme as its workhorse vaccine could no longer be used in about a quarter of the population. What should we make of this outcome?

First of all, we should note that this issue was not picked up in the clinical trials of the AstraZeneca vaccine – the clots were too rare. It wasn't until tens of millions of people had been vaccinated that the problem was detected.

Second, it was only detected even then because this kind of blood clot is extremely rare anyway. If it were a common condition then a link with the vaccine would be much harder to detect – the signal would be lost amidst the background noise.

Third, this particular safety issue wasn't anticipated – there was no *a priori* reason for a vaccine to cause blood clots and thrombocytopenia. This meant that the regulators were – reasonably enough – slower to accept a link than they had been with earlier instances of much more predictable allergic reactions.

Fourth, the risk – even if a link were eventually proven – was extremely low.

Statisticians quickly came up with comparisons to put the risk in perspective. It was said to be about the same as the risks of taking a two-hundred-mile car journey or a much shorter cycle ride. A pertinent comparison was the similar risk of dying from deep vein thrombosis linked to a single long-haul flight. Such comparisons are helpful but – as I said in Chapter 7 – perception of risk is a personal matter. Happily, safety concerns didn't appear to have any immediate impact on uptake among older adults and so the vaccine roll-out continued apace.

There were striking differences between the way the risk from the AstraZeneca vaccine were communicated in 2021 and the way the risk from novel coronavirus had been communicated a year earlier. The discussion of the risk from the vaccines was largely unobjectionable, there was a serious and measured approach to communicating the risk to inform individual decision making. The risks were not downplayed but they were put in context.

This was highly commendable, but what a contrast. As I've explained, the risks to the majority of the population from novel coronavirus were consistently and deliberately overstated by both the government and sections of the media in order to bolster support for lockdown. If people could be trusted with the facts about the AstraZeneca vaccine then surely they could have been trusted with the facts about novel coronavirus too.

CHAPTER 14

THE LAST LOCKDOWN?

The third lockdown in England (second in Scotland) began on January 6th 2021. Once again, most of the UK's population – including schoolchildren – were told to stay at home. There was no denying the public health crisis; in the week prior to lockdown being imposed there had been more than four hundred thousand cases. Novel coronavirus-related deaths rose to more than a thousand per day towards the end of the month. Hospitals were under enormous pressure with around forty thousand Covid-19 patients – 80% more than at the April 2020 peak – and more than nine out of ten beds occupied. More of these patients were surviving – a triumph of improved patient care – but that also meant that they stayed longer in hospital to recover.

Dates versus data
Again, the UK had acted late – the lockdown came more than three weeks after the initial announcement that the more transmissible alpha variant was circulating. If we'd responded more quickly we could have managed with less stringent restrictions. My impression is that during those crucial three weeks the public health discussions tangled up the threat of the alpha variant with concerns over social mixing at Christmas. As the

spread of alpha across Europe in the following months demonstrated, it was the variant that was the problem, not Christmas. Yet again, we had our eyes on the wrong ball.

The good news was that numbers of cases declined quickly – by the end of January they had fallen by half. Hospitalisations and deaths lagged behind as always, but were soon declining even faster. On closer inspection the picture wasn't quite so rosy, however. Not all variants were declining at the same rate; cases due to the alpha variant were hardly declining at all. Even during lockdown, the R number for this variant was only just below one. Within a month all other variants had declined to low levels, leaving alpha to dominate. Thankfully, there were already indications that the vaccines were effective against the alpha variant, greatly reducing the likelihood of hospitalisation and death.

The vaccination roll-out was going well too. By February 12th more than 95% of everyone over seventy years old had had their first dose and the UK administrations were starting to contemplate the exit from lockdown. Road maps for the phased lifting of restrictions were laid out.

The mantra of this period was 'data not dates'. For England, four tests were devised that would need to be passed to progress to the next stage of release from lockdown. These were: vaccine roll-out proceeds well; evidence that the vaccines work; no risk of the NHS being overwhelmed as infection rates rise; and new variants do not change the risk assessment.

I welcomed the emphasis on data not dates, but it turned out to be a one-way relationship. I suggested to a House of Commons Select Committee on February 17th that the data on both virus and vaccine were not merely encouraging, they were far better than anyone had been expecting when the lockdown began. If we were being guided by data not dates then surely we should be looking at bringing unlocking forward. Yet schools remained closed and outdoor activities were restricted, even though we knew that children were at extremely low risk and that novel coronavirus does not transmit well outdoors. Why were these restrictions considered necessary?

In the Select Committee session, I illustrated my point about outdoor transmission by referencing the unfounded concerns of the previous summer about people going to the beach, reiterating that there had never been a novel coronavirus outbreak linked to a beach anywhere in the world. This was nothing new – I'd made the same point many times before – so I was completely unprepared for the reaction that followed. My comments about beaches led the TV and radio news bulletins and were on the front pages of many of the newspapers. The story ran for three days. How could something we'd known for almost a year, and supported by a wealth of scientific data, be news?

I wasn't prepared for the hate mail either, as vicious as any I received throughout the pandemic. This didn't make immediate sense to me: if the fact that beaches were safe was news at all then surely it was good news, but evidently some people were not happy to hear it. The best explanation I heard for this kind of reaction was that people who had spent the past year indoors did not want to be told that it had been safe to go out all along.

There was a positive outcome from the furore. My comments were quoted in a parliamentary debate and various experts – including Chris Whitty – backed them up. England's public health messaging was amended to Hands, Face, Space *and* Fresh Air. The BBC News even came up with an animation showing how air flow dispelled virus aerosols. This was all well and good, but it was twelve months overdue. There never had been any need to keep us indoors.

The re-opening of schools was Stage 1 of the road map in

England, but they stayed closed until early March. By that time the epidemic had been in decline for eight weeks, hospital admissions with Covid-19 had fallen by almost 90% and there was no realistic prospect of the NHS being overwhelmed in the immediate future.

I was speaking on Radio 4's *PM* programme when Boris Johnson said in a press briefing that he was concerned there might be a surge in cases as a result of re-opening schools. There had never been a 'surge' in cases associated with schools re-opening, not in the UK nor anywhere else in Europe. Outbreaks, yes; a modest increase in the R number, possibly; but no surge. There was no surge when schools did re-open on March 8th either, perhaps a slight increase in cases that even the statisticians weren't convinced about. My view remains that schools could have been re-opened safely at the end of January, five weeks earlier.

Sadly, the outbreak of common sense over outdoor activities and the overdue re-opening of schools didn't extend to other restrictions. Contrary to their promises, the UK administrations stuck firmly to dates despite ever more positive data. There were two reasons why the UK was being extremely – even excessively – cautious: models and variants.

Reasons to be cautious

The SPI-M modellers – my own team included – had been busy in early 2021. In Chapter 13 I explained our concern that despite the vaccination programme there was still scope for a resurgence of hospitalisations and deaths if the epidemic took off again. This was because the vaccines weren't likely to be 100% protective and because not everyone would take up the offer to get vaccinated. On the other hand, the better the protection and the higher the uptake, the weaker the link between the number of infections and the burden on the NHS.

This is exactly the kind of complex, dynamic situation that mathematical models can help us navigate.

Alongside colleagues from Warwick University and Imperial College, I presented my team's modelling work at a Science Media Centre press briefing in January. Our conclusions were easy to summarise. We were confident that severe cases would continue to fall in the coming weeks. They did. We thought there was a real possibility of a major resurgence of severe cases if all restrictions were lifted in early March. There was general agreement about that, so it was never attempted.

Beyond this, my team concluded that we didn't know what was likely to happen: there was too much uncertainty about everything from the progress of the vaccination programme and the effectiveness of the vaccines to the way people would behave. We therefore declined to estimate the size of any resurgence; it could be large or it could be insignificant, we couldn't say until we had more data. My colleagues were less circumspect and – as usual – the media focused on their worst case scenarios that predicted even bigger waves than the one we'd just experienced.

There was plenty of scepticism in the media about these pessimistic model outputs. Given the genuine uncertainty, the criticism wasn't entirely fair at first, though there was every right to expect that, even if the central predictions were inaccurate, the true course of the epidemic would at least fall within the modelled range of possibilities. As time went by, however, the data began to look better than even the best case scenarios. Once Public Health England's SIREN study had reported evidence of a substantial impact of vaccination on transmission rates – perhaps enough to get us close to the herd immunity threshold – it was hard to see any justification for predictions of a huge wave. My own team's models allowed that there might be no resurgence at all. In the event, there wasn't, at least not for the alpha variant. However, there was still the possibility of new variants.

SPI-M's models of this phase of the UK's epidemic were calibrated for the alpha variant that had been with us since late 2020. Even if the vaccines made alpha much less of a threat, there was always the possibility of new variants that were more transmissible, more pathogenic or more able to evade immunity from prior infection or vaccination, or any combination of these. Since we had no way of knowing in advance the characteristics of a hypothetical new variant, the models couldn't really help, other than to illustrate the self-evident possibility of a wave that could overwhelm the NHS if the variant was problematic enough.

The prospect of a dangerous new variant was used to make the case for continued suppression of the virus, the idea being that if there were fewer cases there was less chance of new variants emerging within our borders. That's far too simplistic – virus evolution is a complicated and inherently unpredictable process and the number of infections is only one consideration of many.

The alpha variant is a good example: phylogenetic analysis indicates that alpha most likely evolved in the UK not at the height of the second wave in early November 2020 but two months earlier when case numbers were still low. I saw the whole argument as nothing more than another poorly conceived justification for extending the lockdown.

In any case, the next new variant was at least as likely to be imported as to evolve locally. The UK government was so concerned about this that they introduced prison sentences of up to ten years for international travellers who failed to declare they had recently visited countries – such as Portugal – deemed to be at high risk of harbouring immune-escape variants. Most commentators thought this grossly excessive. Getting through the pandemic was all about balancing harms, and this policy gave the impression that decision-makers had lost all sense of proportion.

Ironically, just over two months later, Portugal became one of the first countries to be 'green-listed', meaning that travel to and from the UK was permitted and travellers didn't even need to quarantine on arrival. Ten days of hotel quarantine on arrival in the UK was required only for countries on the red list.

Small steps to unlocking

Around this time, the idea of eliminating novel coronavirus from the UK briefly re-surfaced under the banner 'Zero Covid'. Despite the vaccines, the proposal was no more credible than it had been the previous year. The scientific journal *Nature* had just conducted a survey of a hundred experts in the field: the respondents thought that elimination was only a realistic ambition for a select few countries and warned that novel coronavirus was here to stay. Chris Whitty – CMO England – was likewise arguing that we would be living with the virus for the foreseeable future. As the second wave raged across mainland Europe, the Zero Covid campaign faded away when even its most ardent supporters were forced to admit that zero was not a realistic target.

Once the prospect of living with the virus became more widely accepted by policy-makers – even, belatedly, in Scotland – there was renewed interest in measures to augment the vaccination programme, especially by increasing testing. From April onwards, there were several trials of 'fit-to' tests to allow people who tested negative to attend sporting and music events. There was also a trial of test-to-release, the daily testing of contacts of cases as an alternative to self-isolation. A few weeks later, I ordered my first test-on-request from NHS Scotland over the internet. The test kit arrived in a couple of days and I used it before taking my first excursion out of Edinburgh for months.

These were all steps in the right direction but if fit-to tests, test-to-release and test-on-request were good ideas now then they were good ideas when I and a few colleagues had first suggested them almost a year before. The point-of-care testing technology that made these strategies much easier to implement had been available for six months, and over that period we'd endured two lockdowns.

Through April and into May, the combination of SPI-M's models and the threat of new variants discouraged any accelerated removal of lockdown restrictions, despite the ever more encouraging data on both the state of the epidemic and the performance of the vaccination programme. I was interested in what the models would say about the effect of moving to Stage 3 – which allowed more indoor sports, socialising and hospitality – several weeks earlier than planned, but the option of earlier relaxation of restrictions was never considered. It was as though the scientific advisory system, always inclined to discount the harms caused by social distancing, was now effectively ignoring those harms altogether. This was no more sustainable in May 2021 than it had been a year earlier.

Overcaution is not a cost-free option while restrictions do damage to businesses and livelihoods and many people had been assuming that a vaccination programme performing above expectations would translate into an earlier release from lockdown. This perfectly reasonable hope was to be dashed; following a schedule laid out back in February, Stage 3 relaxations only happened on May 17th. By then, novel coronavirus had become a relatively minor public health problem with fewer than a thousand patients in hospital and ten deaths per day, with both numbers still falling. Dates had triumphed over data again.

If we were reluctant to lift restrictions for everyone there was still the possibility of doing so for those who had been vaccinated. A similar idea had been put forward earlier in the pandemic under the guise of so-called immunity passports. At that time the question was whether people who had acquired antibodies to novel coronavirus through natural exposure should be exempted from at least some restrictions. The suggestion was quickly rejected for fear of encouraging some people to deliberately expose themselves to infection.

The idea of immunity passports briefly re-surfaced in November 2020 when Boris Johnson was identified as a contact of a case and required to self-isolate for fourteen days, even though he'd already had the infection. The public health argument was that there was still a chance – however faint – that the Prime Minister might be infected again and go on to infect someone else. This was not a convincing scenario.

My team's models suggested that the vaccine passport approach would work well, accepting there would need to be room for exemptions on clinical advice (as there were for face coverings). This assessment was supported by its adoption in Israel, the only country whose vaccine roll-out was ahead of the UK. The US too removed many restrictions from people who had been vaccinated.

Despite the necessity of vaccine passports for travel to some international destinations, the UK did not go the same route at the time. This was defended as being more equitable, and that may be so, but it meant that millions of people were subjected to restrictions for weeks or months for no discernible public health benefit. There may even have been a public health cost, as an incentive for people to get vaccinated could have helped to maximise uptake.

Up to Stage 3 of the road map, I felt the UK administrations were unnecessarily overcautious. Conversely, as I said on the

BBC's *Andrew Marr Show* in March 2021, I had always been less confident about the planned final stage of the process: the complete removal of all novel coronavirus restrictions in England on June 21st. It wasn't likely we'd have reached the herd immunity threshold by then – it would be at least another month before all adults were fully vaccinated, and even that might not get us there.

This meant that some resurgence of cases was likely if all measures were lifted on June 21st. Hospitalisations and deaths would follow, albeit at a much lower rate thanks to the vaccination programme, but conceivably still enough to cause a significant public health problem. By contrast, in Scotland there was an explicit recognition that a complete return to pre-pandemic normality might not be possible and we would need some countermeasures in place for the foreseeable future.

For these reasons, the final stage of the road map out of the third lockdown was always going to be the trickiest to navigate, but the task became much harder when the delta variant appeared.

Delta

The delta variant was first reported in India where it contributed to a severe second wave in March 2021. As with beta and gamma before it, delta was not recognised as a variant of concern until after it had arrived in the UK in April. That said, delta got a big head start from the sheer volume of imported cases – Public Health England identified hundreds of cases of delta in travellers in the weeks before India was moved to the red list. The subsequent wave of infections could have been delayed by weeks had imported cases been kept to single figures.

The delta variant turned out be substantially more transmissible than alpha. We were now dealing with a virus that was over twice as infectious as the one that had emerged in Wuhan eighteen months earlier, a development that no-one – including me – had anticipated. This corresponds to an increase in R0 (the basic reproduction number) from around three to more than six, a huge change that made novel coronavirus considerably more difficult to control, whatever interventions we deployed.

Surge testing for infection in hotspots such as Bolton and Glasgow slowed local outbreaks, but did not prevent the delta variant spreading throughout the UK and causing a nationwide increase in cases in early June. Much depended on whether or not the vaccines gave effective protection against the delta variant as against alpha. Early data suggested that they were less effective after a single dose but equally effective after two. This prompted increased efforts to encourage people to get their second vaccinations, plus vaccination campaigns targeted at communities where vaccine uptake had been low.

The final unlocking on June 21st was delayed for four weeks for the valid reason that the delta variant had changed the risk assessment and so the situation did not satisfy the fourth of the UK government's four tests. Numbers of hospitalisations and deaths were still low but rising slowly, and numbers of reported cases were rising rapidly – this was a new wave of a new variant and it was difficult to know how big it would be. The delay would allow time for several million more people to receive the all-important second dose of vaccine. The UK government expressed confidence that a four-week wait would make a decisive difference but that wasn't what SPI-M's models – now re-calibrated for the delta variant – showed at the time; the latest batch of predictions were nothing short of alarming.

That's where I shall bring this account of the third lock-down to a close. Delta – like alpha before it – had tipped the balance more in favour of the virus. This latest new variant was inevitably going to cause problems whenever and wherever it appeared, whether it did so when the vaccination programme had barely begun – as in India – or was well under way – as in the UK – or was almost complete – as in Israel. Delta even managed to evade Australia's strict border controls.

The world would battle with the delta variant for many months to come and, once that chapter of the novel coronavirus story had run its course, in all likelihood another variant of one kind or another will appear and a new chapter will begin. We will have to be flexible and responsive to whatever challenges novel coronavirus throws up in the future. Living with the virus will not be easy. We will reach an accommodation eventually – we have no choice – but we were still some way off in June 2021.

In May 2021 the Prime Minister announced that a formal inquiry into the UK government's handling of the novel coronavirus pandemic would begin the following year. One big question for the inquiry is whether the lockdown strategy was justified at all and, if it was, whether it was implemented appropriately. There will always be differing opinions on this. Mine is that the lockdowns were disproportionate and largely avoidable. I'll set out my reasoning in detail later on, but first let's broaden our perspective and examine how other countries dealt with novel coronavirus.

CHAPTER 15

WORLD VIEW

This was a truly global pandemic. By the end of 2020 only a handful of countries – mostly small islands – had not reported any Covid-19 cases. The global death count reached one million in late September 2020, hit two million in January 2021 and three million just three months later. The true numbers will be higher still because of under-reporting – not every death due to novel coronavirus is identified as such.

This is a huge toll, though we need to keep it in perspective. According to the Institute for Health Metrics and Evaluation at the University of Washington, novel coronavirus was responsible for 4% of deaths globally during the first fifteen months of the pandemic, putting it fourth in their cause-of-death rankings, fractionally ahead of other respiratory infections combined. The numbers change to 9% and third in the rankings using estimates corrected for under-reporting, but novel coronavirus remains well behind heart disease and stroke.

Pitfalls of comparing countries

By some estimates, the UK suffered the highest per capita mortality rate due to novel coronavirus in the world during 2020, vying for top spot with the likes of Belgium, Italy, Spain, USA, Mexico and Peru. Rigorous comparisons are tricky because – as we saw in Chapter 5 – tallying deaths due to novel coronavirus is surprisingly difficult and different countries do it in different ways.

Excess deaths show the UK in a slightly better light, though still around the top ten. Of course, the pandemic did not end in 2020 and the rankings may well change over time, not least because the UK's vaccine roll-out was faster than most. Nevertheless, this poor outcome for the UK in the first year of the pandemic inevitably led to comparisons with other countries to try to work out what we did wrong.

This kind of comparison appears deceptively simple and it was certainly popular: the media, politicians, commentators and global health experts all chipped in. I don't object to comparative studies in principle – I use them in my own work – but they need to be done as rigorously as possible and interpreted cautiously. This wasn't the case for many of the comparisons between countries carried out during the pandemic. The result was that despite general agreement that the UK had lessons to learn there was a lot of disagreement on what those lessons might be.

It was easy enough to identify countries that had done 'better' than the UK and also had greater testing capacity, had stricter self-isolation rules, had closed their borders sooner, had made apps compulsory or had gone into lockdown earlier. This was almost invariably interpreted as evidence that if only the UK had done the same then we would have had a better outcome too and was often expressed as a desire that the UK be 'more like' South Korea, Iceland or whatever country happened to support the case being made. This is known as cherry picking. It is poor science and should be resisted.

The problem with these comparisons is that countries differ from each other in a multitude of ways, so identifying the meaningful differences isn't straightforward. There are good arguments that the UK was predisposed to suffer a severe epidemic whatever we did. Here are some suggestions: infection was seeded quickly and widely thanks to our high volume of international travel; our crowded cities and our lifestyles promoted the spread of the virus; and our ageing population and generally poor population health meant that more of us were at risk of developing severe disease. The more these factors contributed to the severity of the UK's novel coronavirus epidemic the less we can blame inadequacies in our response.

I will discuss three countries and one continent that offer different perspectives on novel coronavirus: Taiwan, New Zealand, Sweden and Africa. If I am guilty of my own charge of cherry picking my excuse is that I'm aiming to show that they all illustrate why care and caution are needed for this kind of analysis. There are two kinds of lesson we might learn. There are lessons for the future, such as it helps to have prepared for the right pandemic and to act early when it arrives. Then there are lessons that could have been applied during 2020, such as lockdowns may not be necessary and can do more harm than good.

Taiwan

Taiwan managed the novel coronavirus pandemic successfully with fewer than a thousand cases and just seven deaths by the end of 2020. There were no lockdowns. The huge difference in outcome between Taiwan and the UK is part of a wider pattern. Almost every east and south-east Asian country fared better than almost every European country.

There is an obvious explanation for this dichotomy: SARS. Taiwan and other countries in the region had been concerned about the possible re-emergence of SARS ever since that disease first appeared in 2003. This was reflected in their pandemic preparedness planning. Novel coronavirus is similar to SARS coronavirus – they are closely related – so it is no surprise that these countries were able to respond effectively. The UK and other countries in Europe were much more concerned about pandemic influenza. This left them at a significant disadvantage from which they never fully recovered.

There are many differences between SARS preparedness in Taiwan and pandemic influenza preparedness in the UK. I will highlight two.

First, as we saw in Chapter 1, Taiwan acted extremely quickly, starting at the end of December 2019. We should give full credit to the public health officials in Taiwan for this timely response; I doubt anyone in the UK realised what we were facing that early on. The UK ended up well behind not only Taiwan but also Japan, South Korea and most other Asian countries.

Second, Taiwan responded vigorously, recognising the importance of rigorous case finding, contact tracing, isolation and quarantine immediately. Those are the measures you'd expect to implement if you were prepared for SARS. If – like the UK – you were prepared for influenza your focus would be more on social distancing.

There are three main reasons for taking a different approach to pandemic influenza.

First, most influenza cases are mild with non-specific symptoms, so hard to detect at all.

Second, influenza has a shorter generation time, roughly two to four days. It's four to six days for novel coronavirus. That difference matters because it makes influenza harder to stop using measures such as self-isolation of symptomatic cases and contact tracing – everything has to happen even faster, almost impossibly so.

Third, the maximum R number for influenza was expected to be considerably lower than it was for novel coronavirus, meaning that less stringent – and therefore less harmful – social distancing measures would be enough to control an influenza pandemic.

There are lessons to be learned for the future from the SARS-influenza dichotomy. One that I'd emphasise is not to plan for just one kind of pandemic threat, a point I've been making to governments and international agencies for years. Another is to act fast in the early stages of a pandemic, something I tried hard but failed to make happen in the UK in January 2020.

I am less convinced by the argument of Jeremy Hunt MP and others that the best way for the UK to tackle novel coronavirus was to be more like South Korea. Putting aside the fact that the UK is a quite different place than South Korea, that proposition wasn't made until March 2020 when it was already far too late. If we had been more like South Korea when we were doing our pandemic preparedness planning in the years leading up to 2020 then – I agree – that would have made a difference. If we had ended up being more like Taiwan that would have been better still.

New Zealand

I have mentioned New Zealand several times already as another example of a country – like Taiwan – that fared much better than the UK. Briefly, New Zealand ramped up quarantine of international arrivals during February 2020, closed its borders on March 19th, went into full lockdown on March 25th, waited until levels of infection were close to zero, and came out of lockdown. Surveillance for new cases continued and if there was evidence of community transmission strict local lockdowns were immediately re-imposed. Australia took a similar approach, though they were more frequently forced into lockdowns.

New Zealand fared much better than the UK during the

first year of the pandemic. The public health burden was far smaller, just twenty-five deaths. Though there was significant economic damage – particularly to the tourism and higher education sectors – this too was less than in the UK. At the same time, it is not true to say – though some have done so – that New Zealand is a model that the UK could have followed.

The difference is in timing. We're not talking about calendar date – as it happens, New Zealand's first lockdown came two days after the UK's, not before – but timing in relation to the state of the national epidemic. New Zealand's lockdown came before there was widespread community transmission of novel coronavirus in that country. The UK's did not.

Studies of virus genomes show that thousands of infections were imported into the UK in January and February 2020 and community transmission was well-established by March. The majority of those early introductions into the UK came from Spain and France. New Zealand is far less well-connected with those countries.

To achieve the same benefit, the UK would have had to close its borders completely several weeks before New Zealand did so on March 19th. Once the epidemic was firmly established in the UK there was no way back, no route to the near-elimination state achieved by New Zealand. This is apparent from the second waves that countries around the world experienced towards the end of 2020. Every country with a first wave of similar scale – adjusted for population size – to the UK experienced a major second wave. The same is true for every country whose first wave was half, a quarter or even a tenth as bad as the UK's.

Whatever the deficiencies of the UK's pandemic response, our failure to emulate New Zealand wasn't one of them. New Zealand never had to solve the problems the UK and every other badly affected country faced, so it's a meaningless comparison.

Sweden

Now let's switch to a comparison much closer to home: Sweden. The Swedish public health authorities – led by Anders Tegnell, State Epidemiologist – said from the outset that they believed they would be living with novel coronavirus for a long time and their pandemic response had to be sustainable.

Sweden was criticised for adopting a so-called 'herd immunity' strategy. This isn't an accurate description of their policy given that they implemented a range of measures to limit the spread of the virus. As a consequence, they remained well below the herd immunity threshold throughout 2020. Whatever label is attached to their strategy, the Swedish experience in 2020 compared unfavourably with its Scandinavian neighbours. Naturally enough, this led to further criticism of their approach.

The relevant comparison, however, is between Sweden and the UK. In 2020, the two countries had similar epidemics but Sweden imposed markedly less stringent restrictions – well short of the full lockdowns in the UK – and incurred substantially less economic harm as measured by the reduction in GDP. For all but the oldest students, most schools stayed open in Sweden. The Swedish experience is profoundly uncomfortable for those who insisted that a strict lockdown was necessary to suppress the virus.

Whenever this point was raised there was a chorus of objections that Sweden was a completely different country to the UK and it was misleading to compare the two. As a debating point, this would have carried more weight if the objectors hadn't been people who – often at the same time – were campaigning vociferously that the UK should be more like South Korea. Nevertheless, let's look at the arguments that Sweden is different in more detail.

One of the alleged differences between the UK and Sweden is that their demographics differ in ways which might affect the spread of the virus. Though there is some truth in this for the UK as a whole, it is not true for the whole of the UK. So, let's compare Sweden with Scotland.

Sweden has a slightly more – I stress *more* – urban population than Scotland: 87% versus 83% live in urban areas. This matters because novel coronavirus spreads more rapidly in cities. On the other hand, Scotland has a slightly lower proportion of people in one-person households and a slightly larger average household size. This matters because the virus spreads efficiently within households.

It wasn't likely that these modest differences in opposing directions would have had much influence on the course of the two epidemics. To find out if that was correct, my team compared the trajectories of cases and deaths per capita in Scotland and Sweden throughout March and April 2020, including the all-important exponential growth phase before either country had implemented countermeasures. They were almost identical. The most important arbiter – the novel coronavirus itself – did not behave as if there was much difference between the two countries.

This leaves the question as to how Sweden managed to control novel coronavirus without lockdowns unanswered. The rate of decline of the Swedish epidemic in spring and early summer was somewhat slower than in the UK and other European countries but it was achieved without closing most schools (except to the oldest children) or most restaurants and – most important of all – no instruction to stay at home. To me, that is yet more evidence that getting the R number below one can be achieved without a full lockdown.

Detractors of the Swedish approach then resorted to the claim that Swedes were more likely to follow voluntary guidelines and take their own precautions against infection, alleging that the same approach wouldn't work in the UK. If Sweden's avoidance of lockdown was really due to vaguely defined cultural differences then we could try to identify suitable cultural levers to achieve the same result in the UK. Better that than unthinkingly assuming that nothing other than lockdown would work here.

The critics were quick to claim that Sweden's approach had 'failed' when – like every other country in Europe – Sweden experienced a second wave in late 2020. Those claims were misleading: Sweden's second wave was discernibly less severe than the UK's and it was brought under control at much the same time, again without imposing a full lockdown. The next wave – as the alpha variant took off – led to more restrictions, as it did everywhere in Europe, but still no full lockdown.

I take two lessons from Sweden's experience. First, it is possible to control novel coronavirus without imposing a full lockdown. Second, a government is more likely to take that course if they recognise from the outset that the measures they take have to be sustainable. As to whether Sweden did better than the UK, although many Swedes have been critical of their own country's response I have yet to hear of any who would swap their 2020 for ours.

Africa

Africa offers a different perspective on the balance of harms from novel coronavirus. I have worked with fellow scientists in Africa for over thirty years and in 2020 I was the director of a multi-national health research partnership called Tackling Infections to Benefit Africa (our acronym – TIBA – is a Swahili word meaning 'to cure an infection'). Our nine African partners are all leading biomedical research centres. TIBA was heavily involved in the pandemic response both in individual countries and – working closely with the World Health Organization Regional Office in Africa – internationally too.

My last overseas trip before travel restrictions came into force was a visit to TIBA's partners in Sudan in February 2020. My colleague Maowi Mukhtar introduced me to the team leading Sudan's pandemic response. He was keen to emphasise that, although budgets were small and resources were limited, his colleagues had a lot of recent experience with infectious disease outbreaks ranging from cholera to Rift Valley fever.

As we sat in one of Khartoum's endless traffic jams on the way back from that meeting, Maowi and I discussed the likely impact of novel coronavirus. At the time, public health experts and many commentators were predicting a catastrophe as the virus would quickly overwhelm weak health systems. Maowi and I demurred. We both remembered similar concerns about the swine flu pandemic in Africa ten years before that had come to nothing. We didn't think novel coronavirus would be nearly such a big problem as we were being told. We weren't the only ones; in September that year I was on the panel for a World Health Organization African Region press conference alongside a Ugandan colleague who was adamant that the threat of novel coronavirus to Africa had been exaggerated from the outset.

For the most part, our expectations look to have been correct. By the end of the 2020 there had been fewer novel coronavirus-related deaths – less than thirty thousand – reported across the entire sub-Saharan Africa region than there had been in most larger European countries. Around half of the deaths in Africa were in South Africa, the most developed country in the region, but even there the per capita mortality rate due to novel coronavirus was well below the figures for Brazil or Mexico or, for that matter, the UK.

I recall arguing with a reporter from the *New Scientist* who insisted that this must be down to poor surveillance and massive under-reporting. I didn't agree with the implication that huge numbers of unexplained deaths due to a viral pneumonia were being missed in most African cities. Under-reporting was a problem; the Institute for Health Metrics and Evaluation later estimated that an estimated 70% of novel coronavirus-related deaths in Africa went unreported, a little higher than the global estimate of 60%. Even so, correcting for this did not put the public health burden due to novel coronavirus in Africa anywhere near that of other infectious diseases such as malaria, tuberculosis or AIDS.

There were several reasons – even back in February 2020 – to expect that the novel coronavirus pandemic would be less severe in much of Africa.

One was that the environment – particularly the climate, perhaps linked to a more outdoor lifestyle – was less conducive to the spread of the virus, implying a lower R number and smaller epidemics. South Africa, of course, has a more temperate climate and is more urbanised than most other African countries, which is consistent with this explanation.

Another difference is demography. Africa's population is much younger and so a much smaller fraction is at risk of a severe outcome of infection. Simply based on age alone, we'd expect the infection fatality rate in Africa to be around a third of that in western Europe. There was a concern about multi-generational households but, on the other hand, there are few care homes in Africa.

My colleague Gordon Awandare at the University of Ghana had a different idea. He wondered if the generally higher level of infections circulating in African populations meant that people's immune systems were trained not to over-respond to novel coronavirus. That was plausible: a lot of the damage caused to patients is actually inflicted by their own body's attempts to keep the virus in check.

This proposition has echoes of the so-called 'hygiene

hypothesis', the idea that if our immune systems have too little to do then they can malfunction, turning against harmless antigens – leading to allergies – or even self-antigens – leading to auto-immune diseases. The hygiene hypothesis is controversial because it implies that some level of infection is 'good' for us and most public health practitioners would strongly disagree with that.

Gordon's idea was subtler. There's plenty of evidence that different infections can interact with one another in ways that affect the outcome for the patient. These simultaneous infections are called co-infections. Sometimes co-infections make matters worse but sometimes – perhaps by regulating our immune systems – they result in less severe disease. Co-infections might well reduce the pathogenicity of novel coronavirus, in Africa and elsewhere.

Environment, demography and co-infections might work in Africa's favour but in the early stages of the pandemic evidence started to emerge in the UK that people of black African descent were more at risk of severe Covid-19. This was a direct concern to me as we are a mixed-race family – my wife, Francisca Mutapi, is black Zimbabwean (and, incidentally, the first black female professor in the University of Edinburgh's four hundred year history). Having looked at the data, we felt that not all this increased risk was directly attributable to ethnicity. It was at least partly explained by environmental factors correlated with ethnicity such as occupation and housing conditions, and by underlying health conditions that are more prevalent in black people in the UK.

Francisca was working tirelessly on the pandemic response in Zimbabwe. A key issue was the provision of care by an already hard-pressed health service. Intensive care units were few and far between and there was little prospect of increasing capacity quickly. So she took another tack. We did some work with Meghan Perry – an infectious disease consultant at the Western General Hospital in Edinburgh – demonstrating that the biggest life saver was the provision of oxygen to patients who needed it. Francisca set about persuading the Zimbabwean government to invest in a new oxygen production facility – a far more realistic goal than installing large numbers of intensive care units. Globally, this issue didn't receive the attention it deserved until the spring of 2021 when a severe shortage of oxygen in India cost many lives.

Another concern was access to PPE. Francisca's colleagues in Zimbabwe did a rapid study of novel coronavirus exposure in hospitals and found that the non-clinical staff were just as much at risk as doctors and nurses. They needed face masks urgently and Francisca procured a substantial donation from a large pharmaceutical company to fill the gap in supply.

Zimbabwe and most (but not all) other African countries took the threat from novel coronavirus very seriously. The majority imposed lockdowns at one time or another. Some were highly restrictive, some less so; some were strictly enforced, some barely enforced at all. From the outset there was concern that these lockdowns would cause harms out of all proportion to the threat posed by the virus.

Organisations such as the World Bank, UNICEF and Save the Children published a steady stream of reports forecasting huge increases in poverty, malnutrition and starvation, together with health crises caused by interruptions to the delivery of childhood immunisations and malaria control programmes. Educational harms would be huge and closing schools would have wider consequences as well, given that many health and welfare programmes in Africa are centred on schools. They anticipated that altogether these would result in far more deaths than would be caused by novel coronavirus.

We can and we will debate whether lockdown caused more

harm than novel coronavirus in the UK and other western European countries. I don't think there'll be much debate about that issue in Africa though. For every country in sub-Saharan Africa – with the possible exception of South Africa – there can be little question that the cure will turn out to be far worse than disease. That's a lesson for everyone to learn, not least the international agencies directing the global pandemic response.

World Health Organization

The World Health Organization has been a positive force in global health since it was founded in 1948. During the novel coronavirus pandemic, the World Health Organization was – as you'd expect – widely seen as the most authoritative source of information on novel coronavirus and of guidance on measures that countries should take to control its spread. For the most part, it fulfilled that role well, accepting that its advice changed as we learned more about the new virus – that was inevitable.

Sadly, though, the World Health Organization got the biggest calls completely wrong in 2020. I'll talk about three issues in particular: the slow acknowledgement of the severity of the threat; the failure to recommend an international travel ban when it could still have made a difference; and the promotion of lockdowns long after it became clear that in many places they were causing more harm than the novel coronavirus itself.

Let's start with the delays to acknowledging the threat. The World Health Organization has two main levers at its disposal.

The first is to declare a public health emergency of international concern, which it only did on January 30th, though novel coronavirus had been detected outside China seventeen days earlier.

The second lever is to declare a pandemic, which it did on March 11th, weeks after it was apparent to most epidemiologists that a pandemic was well under way. It matters that these declarations were made far too late because they are necessary precursors for what the World Health Organization calls 'co-ordinated international action'.

One co-ordinated action that could have made a difference right at the beginning of the pandemic was a comprehensive ban on international travel. This is not hindsight. Ever since the 2003 SARS epidemic, epidemiologists have been well aware of how fast a new virus can spread through international travel, especially air travel. We knew that to have any chance of averting a pandemic action must be taken very quickly indeed. Ideally (from an epidemiological perspective), China would have closed its borders completely in early January 2020.

Next best would be for the World Health Organization to have acted on January 13th when the first novel coronavirus case was confirmed outside China, in Thailand. That was never likely though. As late as February 29th they declared that *the World Health Organization continues to advise against the application of travel or trade restrictions to countries experiencing Covid-19 outbreaks.* The only redeeming feature of this declaration is that by the end of February it was already too late for a travel ban to make a decisive difference in many countries.

It wasn't only the World Organisation who was fault. I should point out that few, if any, UK scientists were calling for the border to be closed at the outset of the pandemic. I didn't mention border closures in my January 2020 e-mails to the CMO Scotland. Why not? My less than satisfactory explanation is that I didn't think it could or would be done – it was too big a step, too quickly, and the suggestion would have been seen as massively disproportionate. I was almost certainly right about that but it doesn't get me – or anyone else – off the hook. It still would have been correct advice and I didn't give it. Looking back, I made a judgement about the credibility of the advice I was providing that wasn't mine to make.

We will need to work out how to persuade the world to suspend international travel immediately should we find ourselves in a similar position in the future. Such drastic action wouldn't be warranted every time we find a new human virus. As we shall see later, that happens almost every year, and most of them do not cause pandemics. Yet it would have helped enormously had it been done early in 2020. Even if a global travel ban wasn't implemented quickly enough to prevent a pandemic entirely, it would have slowed it down, stopped the virus being seeded so widely, and bought the world some time.

I explained the World Health Organization's rationale for recommending lockdowns in Chapter 4. To recap, lockdown was conceived as a time-limited, localised intervention with the aim of eradicating novel coronavirus entirely. It was never intended to be a sustainable intervention implemented on a global scale. It should have been abandoned once it became apparent that eradication was not feasible (with the exception of countries that were in a position to try for elimination – and to completely close their borders).

Yet the World Health Organization continued to promote lockdown even as cases and deaths soared all over the world. The lockdown strategy clearly wasn't working and was causing immense harm at the same time – the worst possible outcome. The key was to recognise much earlier on that novel coronavirus was not going to be eradicated and to promote sustainable responses to the pandemic. The World Health Organization did finally come round to this view, but not until October 2020. Even then, it never spelled out what an alternative to lockdown might look like for countries unwilling or unable to inflict such a damaging policy on their people.

It remains to be seen where all this leaves the World Health Organization in the long run. No doubt there will be calls for it to be given more powers and to be more agile, and also calls for it to be reformed or even replaced. A report by one high-level body – the Independent Panel for Pandemic Preparedness and Response – published in May 2021 mentioned all those options without making a recommendation, but did acknowledge that the early global response to the pandemic was woefully inadequate.

Whatever the World Health Organization's fate, we need firmer international commitments to underpin a co-ordinated international response to the next pandemic, as we do to other global health challenges such as antimicrobial resistance. Experience of climate change – another problem that cannot be solved unless countries act together – shows this will not be easy. It has to happen though.

CHAPTER 16

SAGE SCIENCE

The fact that the UK had one of the worst novel coronavirus epidemics in the world sits uncomfortably with the widely held view that the UK is good at science. Using any measure of scientific achievement – numbers of research papers published, numbers of times those papers are cited by other scientists, number of Nobel Laureates – the UK comes out as punching well above its weight. That strong science base should have given us an advantage during the pandemic, yet we fared so badly.

Role of advisors

The UK also had the advantage of a well-developed system for feeding scientific advice into government policy. This process is the responsibility of scientific advisors appointed by all key government departments and by the devolved administrations. The most senior of the network of advisors are the Government Chief Scientific Advisor (CSA) and the Chief Medical Officer (CMO) for England.

In 2020 the CSA was Patrick Vallance and the CMO was Chris Whitty. The CSA chairs the Scientific Advisory Group for Emergencies, usually known as SAGE. SAGE is convened to deal with crises such as flooding, nuclear accidents or pandemics. It first met to discuss novel coronavirus on January 22nd 2020 and continued to meet at least weekly throughout the year.

SAGE handpicks many of its members according to the nature of the crisis before it. The 2020 membership was extremely well qualified and brought a wealth of expertise and experience of infectious diseases to the table. The same was true of the SAGE sub-committees: SPI-M, which advises on epidemiological modelling; SPI-B, which advises on behavioural science; and NERVTAG, which advises on viral threats.

The majority of scientific advisors are unpaid. The CMOs and CSAs are on the government payroll but the rank-and-file are seconded, pro bono, from their regular jobs, be these in universities or government agencies. Even so, as you'd expect, advisory committee work became the main activity for many of us in 2020–21, often eating up every working hour and more.

I know many of the scientists on these committees well and have the utmost confidence in them. I know the quality of the work they do. The advisory system gave the UK government direct access to some of the best expertise in the world. Yet, somehow, the UK still ended up with one of the world's worst novel coronavirus epidemics. So what happened? Did the scientists give good advice, which the government ignored? Or did the scientists give bad advice, which the government acted upon? There are no clear-cut answers to those questions but it seems to me that there were faults on both sides.

One of maxims within government is: 'advisors advise, ministers decide'. In some ways it's fair comment; most of the time scientists play a limited role in policy development. One SAGE member expressed those limitations in saying: 'the role of the scientific advisor is to help the government do things ever so slightly less wrong'. The problem with this is that it lets the advisors off the hook by minimising their responsibility for the actions that are taken on the basis of their advice.

On the other hand, during the pandemic we often heard ministers say that they were following the science. Ministers have used that line for as long as I can remember, and I've always been wary of it. It lets ministers off the hook by hiding behind the scientists. The truth is somewhere in between. Greg Clark MP – Chair of the House of Commons Science & Technology Select Committee – captured this when he said to Chris Whitty that ministers are not being given much leeway when the government's own Chief Medical Officer tells them that if they didn't impose a lockdown immediately the NHS would be overwhelmed. It was a fair point.

I have advised the UK government on infectious diseases for more than twenty years, covering a variety of difficult topics including pandemic influenza, foot-and-mouth disease and bovine tuberculosis. My aim is always to explain as best I can the pros and cons of the options on the table, not to advocate for one or the other. As I said on Radio 4's *Today* programme one morning in June 2020, I felt that I had done my job as an advisor if I was heard, and if I was understood. That's all.

Not everyone saw it that way. It became routine during the pandemic for scientific advisors and commentators to push for one policy or another. As citizens they have every right to do that, but when there is a constant stream of government scientific advisors in the media calling for a lockdown – as there was during the second wave in October 2020 – then we have crossed a line. This is not advice, this is advocacy.

Such behaviour would be an issue even if we were dealing with a purely public health problem, but we were not. The ramifications of lockdown extend far beyond any short-term public health benefit. Yet, as Patrick Vallance has openly admitted, SAGE was never tasked with the bigger picture, its primary role was to advise on the controlling the epidemic. It was exceptionally well qualified to do that but – given that so many of its members were clinicians and public health experts – its advice was always going to err on the side of averting an immediate health threat with too little emphasis on long-term harms.

Several explanations have been put forward as to why the UK's science advisory system wasn't more successful in 2020: group-think, unconscious bias, tunnel vision, hubris, discouragement of dissent and lack of diversity, to name a few. I accept each of these could have played a part, though I'd add that most of these charges could be levelled at almost every committee I have ever known. There's no question that everyone involved was doing their best.

The question that bothered me more was whether the government was using its own advisory system to best effect, whether the advisors were being asked the right questions. Government seemed to spend much more time thinking about tactics – such as how many people could meet or what times pubs had to close – than about strategy. Even senior ministers opined on minutiae such as the maximum length of a queue for takeaways, a question surely best left to the public health agencies. Strategic questions were left unasked and unanswered.

For example, SPI-M was never asked to model alternatives to lockdown. Most of the work on identifying alternatives was done outside not inside the science advisory system. That's hard to justify – it's not as if lockdown was uncontroversial, so why not use the huge scientific resource available to government to try to find another way?

A sub-optimal relationship between science, scientific advice and policy was an unexpected weakness given that the UK had invested in both pandemic preparedness planning and a sophisticated system for communicating scientific advice to ministers. This wasn't the only line of communication, however. The novel coronavirus epidemic was the biggest news story of the year by far and there were plenty of voices eager to participate in the discussion.

Any number of scientists – not least many of us on the advisory committees – were vocal in the media (usually 'speaking in a personal capacity' – a phrase that some found irritating but signified only that they were not hearing an official pronouncement by any government department or agency). Throughout the year there was an almost daily public airing of topics such as the necessity of lockdown, the pros and cons of keeping schools open, what relaxations could be permitted at Christmas and who should be the first to be vaccinated.

I think it was right that this happened. Exactly the same debates were happening in research seminars and advisory committee meetings, and the issues being debated there had too big an impact on too many lives to not be discussed freely and openly. The public discourse kept pace with the scientific discourse pretty well.

Sometimes the scientists spoke in more co-ordinated fashion. These included well-respected bodies such as the Royal Society of London and the Academy of Medical Sciences. There were also pop-up groupings such as Independent SAGE – led by ex-CSA David King, who I'd worked with on several occasions since the foot-and-mouth disease epidemic of 2001 – and lobby groups such as the Tony Blair Institute for Global Change. All had useful contributions to make, though they didn't always agree with government policy.

I suspect that the net result of having so many competing voices was that it became easier for politicians and policy-makers to ignore all of them. That said, I am strongly of the view that every voice has the right to be heard, taking the view attributed to Voltaire by his biographer Evelyn Hall: 'I may not agree with what you say but I'll defend to the death your right to say it'. I must add the standard rider that with freedom comes responsibility. That's true of scientists as much as anyone.

Doing research at pace

There was certainly a lot of science to explain, comment and advise on. The pace of scientific research on novel coronavirus during 2020 was extraordinary and unprecedented. Thousands of research groups around the world – including my own – were re-purposed to work on the pandemic threat. Tens of thousands of scientific papers were written on every conceivable aspect of the biology of the virus and all sorts of ideas for tackling the pandemic.

The volume of research output was far too great for any one individual to keep track. The Usher Institute – where I'm based at the University of Edinburgh – quickly set up teams to carry out rapid reviews and syntheses of the current evidence on pretty much any topic relating to novel coronavirus. This was an unheralded contribution but a valuable one and I hope it will become standard practice for major scientific issues.

In normal circumstances, the process of translating scientific knowledge into policy is painfully slow. The link between smoking and lung cancer was first identified in 1950 but it took decades before serious efforts were made to reduce smoking.

There is a well-defined sequence of events: the research is done; it undergoes peer review; it is published in a scientific journal; more research is done that replicates or refutes the findings; a consensus is reached among scientists; this is conveyed to policy-makers and fed into the policy-making process. At every stage the science is subject to scrutiny and challenge, and new evidence may emerge that shifts the consensus. Entire books have been devoted to unpicking how this process works.

During the pandemic the time between doing the research

and informing policy – normally years or decades – was reduced to weeks or even days. Scientific advice had to be based on whatever relevant information was available at the time and sometimes policy decisions had to be made with a minimum of evidence, but there was no way around this. I recall explaining to a Scottish Government minister that it was entirely possible the advice that my colleagues and I were giving could change as new evidence emerged. The minister completely accepted this but pointed out that it was difficult for politicians if they then had to change policy as a result – they would be accused of U-turns and flip-flopping. Following the science wasn't going to be as easy as it sounded.

It is right to ask whether the pace of work meant that there was too little quality control on the research being done and that resulted in poor scientific advice to government. I don't think it did. I am not aware of any key decision being made on the basis of scientific evidence that hadn't been scrutinised by one committee or another. In any case, the question the policy-makers should be asking is not what does any individual scientist or scientific paper say but what is the consensus view. A strength of the UK's advisory system is that it aims to look at all the available evidence and reach consensus, while acknowledging disagreement where disagreement exists.

In my view, the bigger problem was that shifting consensus can be a slow process. I felt this most acutely when it came to schools. My reading of the evidence from as early as March 2020 was that it was safe to keep schools open, but it took much longer for a majority of my colleagues to be convinced. I have mixed feelings on this. Of course it was a disaster that so much schooling for millions of children was lost. However, it would have been wrong for policy to change on the basis of a minority scientific opinion. When the evidence became overwhelming, the consensus changed.

I think the lesson here is more about prejudgement. The assumption that children were virus factories and schools would drive the pandemic – taken straight from the influenza chapter of the epidemiology text books – was so ingrained that it took a greater weight of evidence than it should have to overturn it.

As I've already mentioned, a crucial step in the scientific process is the publication of research findings following peer review. Like most working scientists, I have often taken issue with peer review – especially when a cherished paper is rejected by a leading journal – but it can be a positive and constructive process and, overall, my own publications have been the better for it. In any case, despite numerous attempts, no-one has yet come up with a better system for the quality control of research outputs.

Peer review isn't perfect though. Not only do good papers get rejected but, just as worrying, bad papers sometimes get published, such as the retracted paper on the safety of the MMR vaccine in *The Lancet* that I mentioned in Chapter 3. For that reason, I have always argued that peer-reviewed publication in a scientific journal is not enough by itself. Further corroboration is needed before scientific evidence is considered robust enough to inform policy.

In 2020, peer review was largely bypassed anyway. The process of publication following peer review was simply too slow. In normal circumstances it takes several months and – even if it were accelerated for particularly important papers – there would still be a delay of weeks before publication.

The solution was the publication of pre-prints, advance copies released prior to peer review. This quickly became the norm and novel coronavirus research was sometimes described as science by pre-print. Many scientists complained that if the paper hasn't been through the peer review process we had no way of judging its quality. Actually, we did: we could make that judgement for ourselves, something we should be doing anyway if drawing on the work to inform policy.

In my view, science by pre-print worked surprisingly well in 2020. I'd go so far as to say that in many respects it worked better than the conventional route of publication in scientific journals. I have already mentioned the pros and cons of the peer review process but the biggest problem with scientific journals is the editors.

Journal editors have immense power over what research gets published yet they are not accountable to the scientists whose work they judge, nor to the users of that research, such as government. The vast majority of journal editors are well meaning and good at their jobs, but they are not all knowing and they are no more free from bias – conscious or unconscious – than the rest of us.

Some journals dealt with this better than others, openly acknowledging the problem. I was particularly struck by a blog written to accompany a paper on modelling herd immunity published by one of the world's leading journals called, simply, *Science*. In the blog, the journal's editors said that they were concerned that the paper gave a different result to other modelling studies and might be interpreted as downplaying the severity of the pandemic. Nonetheless, the paper had passed peer review and they felt that it was in the public interest to publish.

Science was more enlightened than most; some other leading journals seemed to favour certain perspectives on the pandemic over others. Science by pre-print lets us see all the novel coronavirus research there is, not a subset of it filtered by editorial whim.

Most people, of course, don't read scientific journals – their understanding of the science is filtered through the media. I had countless conversations with journalists in 2020 and – putting aside the inevitable desire for a punchy headline – my experience was that the great majority wanted to get the facts right, even if they had their own slant on the story.

I recall talking to a *Mail on Sunday* journalist one Saturday morning about a specific entry in a table in a scientific paper, purely to confirm whether or not the data were consistent with an article the paper was preparing on novel coronavirus in schools. That's impressively diligent journalism.

In this book so far, I have been critical of BBC television's coverage of the pandemic – their main news programmes repeatedly misrepresented the risks from novel coronavirus – but it wasn't all bad. If you dig into the health and science sections of the BBC News website James Gallagher, Nick Triggle and colleagues were providing informed and more balanced coverage.

The media's efforts to explain the science of novel coronavirus and its implications were greatly helped by the work of the Science Media Centre. The Centre was set up in 2002 with the remit of improving communications between scientists and journalists. This was the era that Ben Goldacre described in his 2008 book *Bad Science*, characterised by alarmist reporting of threats to our well-being and over-hyped claims for possible cures and remedies. The Centre's aim was to deliver more accurate and better balanced science reporting and it came into its own during the novel coronavirus pandemic.

The Science Media Centre's way of working is to highlight the latest scientific reports and publications and post comments from those on its extensive list of scientific experts. These posts are often widely picked up by journalists and news agencies in the UK and beyond. The Centre often links individual scientists with individual reporters, and regularly holds briefings for the media on hot topics – I spoke at several of these myself.

The Science Media Centre provided journalists with an accessible source of authoritative scientific comment on the epidemic and its ramifications which was used to full effect, and we all got better quality media coverage as a result.

Science that made a difference

The UK was a major contributor to the global output of research on novel coronavirus – as I said, we have a strong science base. The UK spends tens of billions of pounds on scientific research every year, through government, industry and charities such as the Wellcome Trust. This huge amount of money is distributed to many different pots. I have wondered for years whether the right amount of money was going into the right pots. The novel coronavirus pandemic convinced me that it was not.

The Rolls-Royce disciplines of biomedical science are molecular biology, cell biology, immunology, physiology and the 'omics' (a collection of high-tech approaches including genomics, transcriptomics, proteomics and metabolomics). These are the disciplines that have the highest status within the scientific hierarchy, get the most investment and the newest buildings, publish in the top journals and receive the most prestigious prizes.

Much of the work done in these Rolls-Royce disciplines is not immediately practical and is described as 'fundamental' science. As such, it is aggressively defended whenever there is any suggestion that science budgets should reflect utility as well as the more nebulous concept of 'excellence'. A lot of fundamental biomedical science is done using animal models and I once had a head of department who regularly called by my office waving the latest cover of a leading science journal exclaiming, 'Look Mark, another great advance for mousekind!'

Towards the end of 2020 I was reflecting on the contribution of the Rolls-Royce disciplines to the pandemic response so far. The answer I came up with was: practically nothing. It is true that a huge amount of research was being done in these subjects, and it is also true that fundamental science of this kind provided the underpinning for the vaccines developed in 2020. Also, as we've seen, genomics did start to make an important contribution in December. So I am not arguing they are worthless, far from it. I am simply pointing out – politely but firmly – that science's most prestigious disciplines had minimal impact on the ground during the first year of the pandemic.

The disciplines that saved lives during 2020 were ones much further down the scientific pecking order: clinical medicine (mostly patient care), behavioural science, epidemiology and public health. These are less Rolls-Royces and more battered old Land Rovers, lacking both funding and prestige. The interventions that really made a difference – improved patient survival, contract tracing and outbreak investigation, identification of risk factors and advice on hygiene, face coverings and ventilation – were all delivered by these disciplines, many of them at a remarkably early stage. That's how science saved lives in 2020.

I can illustrate this by referencing the research work done on the most important risk factor for severe illness due to novel coronavirus, age. Clinicians and epidemiologists recognised age as a risk factor before the UK epidemic had even begun. This was vital information and allowed patient management protocols and public health interventions to be tailored accordingly. The knowledge that age is a risk factor undoubtedly saved lives. It also showed that it was safe to keep schools open. Within six months, epidemiologists had measured other risk factors so precisely that we could identify 5% of the population who were over fifty times more likely to die from novel coronavirus infection than the other 95%.

It would have helped to know precisely why age and other risk factors were important because then we might be able to identify those most at risk more reliably. There were several contenders: senescence of the immune system, expression of specific molecules on cell surfaces, or more recent exposure to other infections. Yet, despite decades of enormous investment in fundamental research on ageing, cell physiology and immunology, we still don't know the answer. The Land Rovers delivered; the Rolls-Royces never got out of the garage.

I can make a similar case for intervention technologies, meaning vaccines, drugs and diagnostics. New vaccines and therapies were always going to be our main hope for finding a way to live with novel coronavirus. The efforts during 2020 that went into delivering them as quickly as possible were truly remarkable. Vaccine and drug development, and all the fundamental science that underpins those endeavours, have been worth every penny invested in them.

All of that is true but it still misses one crucial point: for the first year of the pandemic, vaccines (outside clinical trials) didn't save a single life in the UK – we didn't have any (though Russia and China were a little quicker). Drugs performed better. Dexamethasone, for example, significantly improved survival of severe cases in hospitals. Its effectiveness in treating Covid-19 cases was demonstrated through a fast-track clinical trial called Recovery – another terrific example of good, practical, high-quality science done at extraordinary pace. Even so, Dexamethasone is not a new drug. Completely novel therapies, such as plasma containing antibodies against novel coronavirus, were less successful.

Contrast this with diagnostics. The RT-PCR test was available within weeks and used around the world. Without it, we would have been working in the dark, case finding and outbreak investigation would have been severely compromised. A few months later, cheap, rapid diagnostic tests had been developed and mass testing on a huge scale became feasible, opening up new possibilities for reducing individual risk and suppressing the virus.

As an epidemiologist, I have always seen diagnostics as a crucial public health tool yet diagnostic development receives nothing like the prioritisation and investment given to vaccines and drugs. Back in 2006, I co-authored a major UK government report – called a Foresight study – on the importance of diagnostics for infectious diseases. The report had a modest impact on investment in the field, but it didn't stop one of the most senior medical scientists in the country telling me that diagnostics weren't cures and so weren't that important. Novel coronavirus proved such attitudes wrong. Of course a RT-PCR test isn't a cure, but when it is used to trigger self-isolation and contact tracing it saves lives.

When we are faced with this kind of crisis again – and I'll say more about that later – I hope that we will have learnt the lessons that 2020 so painfully taught us. A lot can happen while we are waiting for a new drug or a new vaccine to solve the problem, many people can die and a horrendous amount of damage can be done. During that period we will be relying once more on medics learning how to treat patients better, epidemiologists learning how the disease spreads, public health practitioners working out what interventions are most effective and developers of diagnostics giving them all the tools they need to do their work. I hope that in whatever time we have before the next pandemic we will have nurtured those disciplines so that they are there, ready to help when needed.

CHAPTER 17

WHAT SHOULD HAVE HAPPENED

The July 2020 article I wrote for the *New Statesman* concluded with the words *I fear that history will judge lockdown as a monumental mistake on a truly global scale.* So I suppose I shouldn't complain that I was introduced on the BBC's *Andrew Marr Show* two months later as a lockdown sceptic. I don't like the term though; it has a ring of 'climate change denier' or 'flat earther' about it.

The case against lockdown

My lockdown scepticism is rooted in the long list of harms caused by the lockdowns implemented in the UK, the unintended but predictable consequences of trying to control novel coronavirus by shutting down society. My hope – and my main motivation for writing this book – is that lockdown scepticism will become the mainstream view.

I must repeat that I am not playing down the threat of novel coronavirus. For many people Covid-19 is a dangerous disease, though the majority of the population is at relatively low risk. The pandemic always warranted a robust public health response, nationally and internationally. For me, that's not the issue; the issue is whether that response had to be lockdown. I have three reasons to claim that lockdown was an over-reaction to novel coronavirus.

First, we didn't take adequate account of the damage that lockdown would do to lives and livelihoods. We will be paying the price for a generation or more.

Second, the public health benefits of lockdown were over-estimated, skewing the argument in favour of that course of action, however damaging it was.

Third, there were much less damaging ways to save lives and protect the NHS. Those alternatives are the subject of this chapter.

It goes without saying the best way to avoid lockdown is not to have an epidemic in the first place. I have already talked about two courses of action that might have made that ideal outcome possible. We could have implemented border controls and international travel bans right across the world in January 2020. That didn't happen. Then the UK could have taken those same actions unilaterally in February. That didn't happen either. I don't think either option was realistic at the time, too many decision-makers had yet to accept the seriousness of the situation. Taiwan and several other countries were much quicker off the mark. That's a crucial lesson to learn: we will need to make better decisions faster should we find ourselves in similar circumstances in the future.

Having said those actions should have been taken, I must add a crucial rider: it's far from certain how well they would have worked. Remember that border controls delay but do not prevent the importation of cases, so we'd face a continual threat. Even if the UK had successfully contained every novel coronavirus outbreak we experienced, I don't for a moment imagine that epidemics could have been prevented in every single country in Europe, the Americas, the Middle East, South

Asia and Africa. Perhaps more countries would have been able to sit out the pandemic behind closed borders in the manner of Australia and New Zealand – with a lower death toll as a result – but the virus would still be there, life would be far from normal, and we'd still have to find long-term, sustainable solutions. We'd all have to learn to live with novel coronavirus eventually.

For now though, let's focus on the situation that the UK faced from early March 2020 onwards. Given what we knew at the time, could we have done better? Finding the best path through the pandemic was never going to be easy, but the path we took led to the UK suffering both one of the highest mortality rates in the world and a hugely damaging series of lockdowns. Is it possible that we could have saved lives and kept out of lockdown as well? My answer to that question is an emphatic yes.

Alternative views

There are different perspectives on the way the UK should have responded to the pandemic. We have already encountered the view that no death due to novel coronavirus is acceptable. Taken to its limits this argument says we should lock down immediately, as tightly as possible, and stay there until we have rolled out an effective vaccination programme. No country in the world was prepared to take such an extreme course of action. There has to be a balance between the public health harms due to novel coronavirus and the economic and other harms caused by lockdown.

A high-profile proposal for an alternative to lockdown was published in October 2020 in the form of the Great Barrington Declaration. The declaration was named for the home of a think-tank called the American Institute for Economic Research and was led by a group of US and UK scientists.

The Great Barrington Declaration advocated an approach where the vulnerable are protected but the virus is left to circulate in the remainder of the population until enough people have been infected to reach the herd immunity threshold. According to its website, around fifty thousand academics and clinicians signed up to the declaration. I was not one of them. I thought there were serious problems with the Great Barrington approach.

One problem was that the epidemic envisioned in the Great Barrington Declaration would be far larger than the one the UK experienced in 2020. Even if cases were entirely confined to the less vulnerable majority, tens of millions of people would be affected, many times more than were actually infected that year. By definition, this group is less much likely to suffer severe disease and die, but a small fraction of a very large number is still a large number. This was not a viable strategy: we would still end up with many more people needing to be hospitalised than the NHS could cope with.

The second problem was the absence of a convincing plan for confining the epidemic to the less vulnerable majority in the first place. This would have to happen for however long it took for the herd immunity threshold to be reached, most likely several months. I have argued throughout this book that we could and should have protected those who most needed protecting far better than we did, but there are limits to how well that could be done. If the virus were circulating freely it was inevitable that some infection would spill over into the vulnerable minority and cause a substantial burden of death and disease in that group too.

There was a third problem with the Great Barrington Declaration. Back in October 2020 we did not know whether acquired immunity to novel coronavirus would be effective enough to get us over the herd immunity threshold. It was risky to assume in advance that immunity would be as good as it is for infections such as measles – that could turn out to be over-optimistic. If we failed to reach the herd immunity threshold in a matter of months but rather – as some epidemiological models predicted – suffered repeated waves of infection then the strategy wouldn't work at all.

For a time, the Great Barrington Declaration received a lot of attention from the media and politicians in both the UK and the US. This provoked an unrestrained backlash from medics and scientists who lined up to denounce the declaration as dangerous, unethical and poor science. The criticism was so vehement that several journalists asked me why the scientific community was so unwilling to listen to new ideas. When an alternative approach is rejected that forcibly it creates a climate of 'it's our way or nothing' and that's not a good impression to give. Dogma has no place in science. If discussion of alternatives were to be stifled then we were condemned to lockdown.

Back to basics

From my point of view, the most unfortunate consequence of the Great Barrington Declaration debate was that it hardened the view that 'protecting the vulnerable' was somehow inextricably linked to a herd immunity strategy of one kind or another. It wasn't. It is true – as Jeremy Farrar wrote in *The Times* in September 2020 – that suppressing the virus is incompatible with the build-up of herd immunity. In other words, you can have lockdown or the Great Barrington approach, but you can't have both. That's not the whole picture though. There is a middle ground and I believe that is where the answer lies.

Let's start with the evidence. Even before the UK first went into lockdown in March 2020 we knew that novel coronavirus is far more dangerous to the elderly, the frail and the infirm than it is to healthy young adults and children. I do agree with the

Great Barrington Declaration that this simple, straightforward epidemiological fact should have shaped our response. Instead, the UK administrations continued to act as though everyone was equally at risk. They even actively promoted that misperception to try to bolster acceptance of the lockdown strategy.

Having established who is at risk the next step is to devise ways of protecting them. As we saw in Chapter 4, suppressing the spread of the virus wasn't enough by itself; huge numbers (possibly a majority) of people who died of Covid-19 got infected during periods of lockdown. Lockdown didn't save these people; we needed to do more. During the first wave in the UK all we actually did was tell the vulnerable to shield themselves through self-isolation – which was neither desirable nor sustainable – and, belatedly, increase biosecurity around care homes. In or out of lockdown, these interventions weren't enough. Some have claimed that this proves that protecting the vulnerable doesn't work. That is like saying that fire-fighting doesn't work because we failed to put out a raging inferno by throwing a bucket of water at it. The truth is that we didn't really try to protect the vulnerable.

I will turn to economics to demonstrate how little we tried. It will be a while before we know the full economic costs of the pandemic but the UK reportedly had to borrow an extra three hundred billion pounds in the 2020–21 financial year. By definition, we incurred that debt because of the pandemic so it wasn't an either/or choice between controlling novel coronavirus and sparing the economy; other things being equal, a smaller epidemic is better for the economy. On the other hand, we should not have been willing to pay any price at all to control the virus. It was always possible for the cure to be worse than the disease, so we needed to find the right balance.

Working out how much higher or lower the costs would have been had we taken a different approach is a counterfactual problem. We came across counterfactuals in Chapter 4 when trying to estimate how many people would have died if we hadn't gone into lockdown. That was a job for epidemiologists. Now we have to work out what the economic costs would have been in this alternative scenario. The answer will again depend on how we imagine people – and businesses, banks and governments – would have behaved in a different set of circumstances. That will always be disputable and no doubt economists will be debating the question for years.

I will look at the imbalance in our response in a much more straightforward way. The UK spent an enormous amount of money on its pandemic response. Most of that money was linked to the measures taken – including lockdowns – to suppress the virus. By comparison, how much did we spend on protecting the vulnerable, both in care homes and the wider community? The answer is not much at all.

There are about twenty thousand care homes in the UK and well over thirty thousand people in those care homes died with novel coronavirus in 2020. (For context, it's worth noting that over one hundred and fifty thousand care home residents die in a normal year). According to the DHSC, the UK spent about five billion pounds on care home support. That's two hundred and fifty thousand pounds per care home. Several times as much could have been spent on protecting care homes – enough to greatly improve biosecurity measures – and that would still amount to a tiny fraction of three hundred billion pounds.

That's one part of the challenge, but the majority of vulnerable people don't live in care homes. Across the UK, over two million vulnerable people in the community were advised to shield in 2020 (increased to four million in 2021). There are about seven million more households that have people over seventy years old or who are otherwise at elevated risk. More than forty thousand people in these groups died of novel coronavirus infection in

2020. I haven't been able to find estimates of government spend on shielding. It can't have been much. Those asked to shield were given advice and offered help with their grocery shopping. Other vulnerable people got nothing more than a letter advising them to take extra precautions.

Each of those households would have benefitted from personalised advice on how to implement Covid-safe measures and being provided with the means to do so, together with routine testing of contacts and support for shielders if they needed to self-isolate, and for those they were shielding if that happened. We could have spent several thousand pounds per household – a substantial sum by any standards – on these measures and that would still amount to a small fraction of three hundred billion.

I am particularly struck by the neglect of shielders, especially care-in-the-home workers and informal carers. These people had the responsibility of standing between the people they are caring for and novel coronavirus. For most of 2020, they got minimal recognition and received no help.

All in all, the under-investment in protecting the vulnerable is one of the most baffling failures of the UK's pandemic response. To say we spent a fortune on suppressing novel coronavirus doesn't even begin to cover it. The UK took on more debt than at any time since the Second World War and economists anticipate we will be servicing that debt for generations to come. We spent, in comparison, a trivial amount on care homes and almost nothing on protecting the vulnerable in the community. We should and could have invested in both suppression and protection. We effectively chose just one.

U-turn on protecting the vulnerable

Ideas such as protecting the vulnerable, cocooning (protecting the vulnerable by protecting their shielders) and segmenting (targeting interventions at the appropriate subsets of the population) ought not to have been controversial. They were sensible and proportionate approaches to tackling novel coronavirus. Nonetheless, they were fiercely resisted by scientific advisors who had mistakenly convinced themselves that they were not compatible with suppressing the virus. The irony is that they were all adopted in one form or another at various phases of the epidemic, just not quickly or widely enough.

Once vaccines became available, the attitude towards protecting the vulnerable completely changed – you could call it a U-turn. When the Joint Committee on Vaccination and Immunisation published its priorities in December 2020, care home residents, the elderly and the most vulnerable were right at the top of the list. This was broadly welcomed with no mention of the implication that if protecting these vulnerable groups was the right strategy once we had a vaccine then it was the right strategy all along. Vaccination was the most – probably by far the most – effective way we could protect the vulnerable, but it wasn't the only one.

Cocooning was adopted even earlier by offering health care workers priority access to testing to reduce the risk to their charges. Care home workers were also among the top priority groups for vaccination for the same reason. This too was broadly welcomed yet, at the same time, millions of shielders in the wider community continued to be ignored. If cocooning was the right strategy for care home residents then it was the right strategy for the most vulnerable in the community too.

Again, this changed when vaccines became available. The vaccination priority list included social care workers from the outset and was amended in February 2021 to include anyone in the community who received the carers allowance or was designated as a main carer.

Segmenting was an important part of the UK government's novel coronavirus strategy at the other end of the risk spectrum. Allowing children to go to school while most adults were in lockdown was segmenting. Prioritising the re-opening of universities in the autumn of 2020 was also segmenting. If segmenting was the right strategy for schoolchildren and students then it was the right strategy for other segments of the population too.

The realisation has half-dawned. It is widely accepted that the UK's response to novel coronavirus failed care home residents. We now need to accept that it failed the vulnerable in the community as well. We also need to recognise that we failed those at low risk from novel coronavirus too, in their case by imposing not one but three national lockdowns (two in Scotland) that harmed everybody.

I have stressed throughout this book that lockdown was not and was never intended to be a sustainable intervention. Once the virus was firmly established – as it was in the UK from late February 2020 onwards – lockdown could only defer the problem; it was not a solution. Lockdown is what you do when you have failed to keep the virus under control in other, more sustainable ways.

How to avoid lockdown

Let's review the tools we had available to tackle the novel coronavirus epidemic in 2020 and early 2021. I've already mentioned interventions that protect the vulnerable; these include infection control in hospitals, biosecurity for care homes and support for vulnerable people and their shielders in the community. In addition, there were several ways we could suppress novel coronavirus transmission.

First, there were Covid-safe measures: hygiene, PPE, physical distancing, ventilation and so on.

Next, there was self-isolation of cases and their contacts, which depended on effective case finding.

Finally, as what should have been a last resort, there was a series of increasingly restrictive social distancing measures, culminating in lockdown.

We needed to make the best possible use of every other option available to give ourselves the best chance of avoiding lockdowns in 2020 and 2021. It was a trade-off: the better the alternatives were working, the less need there'd be for any social distancing measures, lockdowns included.

It's hard to argue that the UK did its best to avoid lockdown. Despite a huge investment in testing capacity we never fully got to grips with case finding, so large numbers of infections went undetected. We never managed to ensure full compliance with the need to self-isolate either. We were also slow to recognise that we could negate the need for social distancing by making contacts Covid-safe, and we never fully exploited the potential of testing for ensuring Covid-safety.

One reason why we kept turning to social distancing and lockdowns rather than relying on Covid-safe measures was our attitude to risk. The majority of public health officials I encountered in 2020 were extremely risk averse. That attitude was understandable in the early stages of the pandemic and it was defensible if we believed the crisis would be short-lived, but it was no basis for a sustainable novel coronavirus response. We needed interventions that were proportionate, not designed to mitigate the tiniest health risk imaginable without any regard to the impact on our daily lives.

Even where the risk was real, we learned that it was possible to carry out many activities safely. We saw this demonstrated every time we watched sport on our TV screens. The key was Covid-safe protocols integrated with regular testing. If a rugby club can manage that then I don't see why most workplaces

– including mine, a university – cannot. The approach was eventually extended more widely once point-of-care diagnostics became available, but that could have been done much sooner and it could have gone much further.

Social activities too could have been made safer if everyone had access to test-on-request. Instead of embracing this solution we wasted months – and endured two lockdowns – arguing about how to deal with false positives.

There is precedent for slowing the spread of an infectious disease by making contacts safe. For many years, this has been the advice given by public health agencies to prevent HIV/AIDS. HIV is spread mainly by sexual contacts (though also by sharing drug injection equipment). The major focus of anti-AIDS campaigns is on safe sex and knowing your HIV status – which means getting tested – not on banning sex altogether.

Though we could have done more to make our contacts Covid-safe, facilitate self-isolation and protect the vulnerable, I am not claiming that we could have avoided all need for social distancing in 2020 and 2021. I do think, however, that if social distancing measures were required they could have been targeted to deliver maximum public health benefit for minimum indirect harm. That would mean targeting interventions by infection and immunity status, by age and risk group, by specific location (such as a hospital or care home) and by geography. There's nothing new in that – we did many of these things in 2020 – but, yet again, we failed to turn this line of thinking into a coherent strategy and pursue it vigorously enough.

There is already a name for a strategy of this kind: precision public health. It means delivering the right intervention, at the right time, to the right people. It's the exact opposite of a crude, catch-all intervention such as lockdown. If, as is often said, precision is the future of public health then we took a big step backwards in 2020. There were huge differences across the population both in the harms caused by novel coronavirus and the harms caused by lockdown. This was an epidemic crying out for a precision public health approach and it got the opposite.

As well as thinking about where to target our interventions we also needed to think about when to introduce them. The answer is that when the epidemiological data indicated that additional measures are needed – any measures, not just social distancing – these should have been implemented immediately. The guiding principle for avoiding lockdown is that early action can be less drastic action.

This is not quite the same as the frequently heard argument that the UK should have locked down sooner than we did in March 2020 and again in November. I am arguing for a quicker but also a less drastic response, well short of lockdown but still preventing the NHS from being overwhelmed. The rationale is simple: we don't have to impose harsh restrictions in an attempt to drive down levels of infection if we don't let them rise up in the first place.

The triggers for relaxing measures are the mirror image of the ones for imposing them. We knew that increasing numbers of hospitalisations and deaths were 'baked in' – as it was often expressed – once case numbers start to rise. The same is true – but in reverse – when case numbers started to fall. A cautious policy-maker would want to wait until the trend is clear, but I'd argue that a halving of case numbers would have been evidence enough.

Waiting too long compounds the harms done by the lockdown. You will recall that, thanks to the way exponential decay works, the public health benefit of keeping restrictions in place falls off steeply over time, making long lockdowns much harder to justify than short ones.

My emphasis on removing lockdown restrictions as quickly as possible may sound surprising – after all, it's frequently argued that the UK made a mistake by exiting from the March 2020 lockdown too soon. This became the rationale for the slow and overcautious exit from the January 2021 lockdown. It's not correct though. The problem isn't when you start relaxing restrictions, it's where you end up. If you relax too far then the epidemic will get out of control again.

I carefully didn't say 'too far, *too soon*' in the previous paragraph. Novel coronavirus has no memory and it doesn't reward us for being patient. If the R number is above one then the epidemic will take off, regardless of how long it has taken to reach that point. We saw this play out in the gradual lifting of restrictions after the first wave – being patient then didn't spare us from the second wave. There is some merit in unlocking gradually: it makes it easier to go back if the situation changes and it allows more time for alternative measures to be put in place that reduce or remove the need for social distancing in the future. There is no merit in waiting too long to begin the process.

The question of which social distancing measures can be safely relaxed after a lockdown is perhaps the toughest to answer of all. There was good evidence even at the time of the first lockdown that we could have removed some restrictions – especially those on outdoor activities and on schools – much more quickly than we did.

Nor do I accept that the instruction to stay at home was justified beyond the first six weeks. The great majority of individuals, employers and organisations could have been advised, supported and trusted to take precautions to reduce the risk. After all, many day-to-day activities could be made safe and were: several million people did not stay at home because their work was considered 'essential'. This wouldn't be normal life – particularly with regard to social mixing and contacts between households – but it would have been considerably closer to it, perhaps with voluntary social distancing guidelines replacing legal enforcement.

There's an ideological dimension to any preference for voluntary over enforced social distancing, and this coloured discussions of the topic, even among scientists. Yet there's a practical scientific question to be asked as well. We need to know whether the collective impact of voluntary social distancing would be enough to meet the policy objective of not overwhelming the NHS.

You will recall that this approach worked reasonably well in Sweden, but it was claimed that it wouldn't work in the UK and we didn't try. This was despite polls consistently showing that people were greatly concerned about Covid-19 and willing to change their behaviour to reduce the risk to themselves and others. Indeed, it's hard to imagine that everyone would ignore the state of the pandemic around them. So, although no formal analysis has been done, I suspect that voluntary behavioural change could have had a substantial impact if we'd adopted that approach in the UK. It may not have been enough by itself, but it didn't need to be, there were other measures we could use as well.

By the time of the second and third lockdowns we had all we needed to put in place a more comprehensive and sustainable approach to tackling the novel coronavirus epidemic. What was lacking was proper planning for and urgent investment in alternatives to social distancing (regardless of this being enforced or voluntary). I think the UK would have done that planning and made that investment had we accepted from the outset that the virus was here to stay. We would have realised that sustainability was more important than using lockdown to temporarily drive down the incidence of infection to low levels only to have it rebound again.

Let me sum up what should have happened. Above all, we should have done far more to protect those most vulnerable to novel coronavirus – that would have reduced the mortality rate regardless of whatever else we did. Social distancing measures should have been introduced more quickly when needed and removed more quickly too. There should have been a concerted effort to shift rapidly from mandatory social distancing to Covid-safe measures and self-isolation of cases and contacts as the preferred means of suppressing the virus, coupled with a mass testing programme designed to find cases and make activities safer and augmented by voluntary social distancing based on public health advice.

We could have done all this in 2020 and 2021 – no hindsight required. If we had, the UK would have spent far less time in lockdown and still suffered a substantially lower burden of death and disease. That's why I'm a lockdown sceptic.

Mistakes made and lessons learned

There will be many lessons learned from the novel coronavirus pandemic. Numerous issues will be raised by the inevitable inquiries as we reflect on what went wrong in the UK and internationally. I am most interested in what prevented us from taking those steps that would have both saved lives and prevented lockdowns. As I see it, we made three cardinal errors: we repeatedly succumbed to optimism bias, we fixated on lockdown, and we focused too narrowly on the public health harms caused by the pandemic.

The first error was the failure by government – and some advisors too – to accept the scale, severity and duration of the unfolding crisis. They were warned. If you turn back to Chapter 1 you'll find the advice that I and my colleagues were giving in January 2020. I think you'll agree that our advice was accurate. The one point I overstated was the potential impact on mortality. That's because I was using the World Health Organization's initial estimate of the infection fatality rate which – thankfully for all of us – turned out to be too high. Even so, the mortality rate in the UK had doubled by the end of March 2020. In all other respects, what we said would happen did happen.

The reason we were able to give such accurate advice at such an early stage – less than three weeks after the new virus was first reported – is that this was a textbook pandemic. It progressed over the course of 2020 much as any infectious disease epidemiologist would expect. The science wasn't difficult – my undergraduate classes have worked through these kinds of scenario. The maths was straightforward. The hardest thing to grasp was the sheer scale of the coming crisis.

This reluctance of policy-makers to accept the seriousness of the situation persisted throughout 2020. When I first warned of the pandemic's likely impact I was told that everything was under control. When I said that we needed to do mass testing on a scale of millions I was told I was being unrealistic. While I was talking about a coming second wave the talk in policy circles – at least in Scotland – was all about the pipe dream of elimination. When I said the vaccine roll-out would not 'end this by Easter' – as some were claiming – I was called a pessimist. When I said that the virus was here to stay I was accused of being fatalistic.

There is a phrase for this kind of attitude: optimism bias. Government didn't plan for the pandemic the scientists said was coming because they gambled that the scientists were wrong. I wish we had been, but we weren't. Government didn't plan for a second wave because they imagined the worst was behind us. It wasn't. Government didn't plan for a sustainable response because they wouldn't accept that the virus is here to stay. It is.

The second error was our fixation on suppressing the virus through lockdown as the 'right' way to deal with the pandemic. There are echoes here of Daniel Kahneman's concept of anchoring. Lockdown was the first approach used to tackle novel coronavirus and most of the world latched onto it. Lockdown only made sense in the context of eradication, but when the World Health Organization was forced to abandon its eradication strategy it did not change course and lockdown quickly became the international norm.

Once that happened it was extremely difficult for any country to take a different path. There is comfort in following the crowd even while it is stampeding in the wrong direction. We wouldn't let go of lockdown even after the evidence of the harm it was causing became so compelling that the World Health Organization itself came to reject it. Governments around the world – the UK included – couldn't change course because to do so would be an admission that they'd taken the wrong path to begin with. Instead, they doubled down and imposed lockdowns again.

From a government perspective, lockdown had big advantages: it didn't require any forward planning, there was no need to build capacity in advance and no direct financial cost. All lockdown took was government decree and a modicum of enforcement. It was a lazy solution to a novel coronavirus epidemic as well as a hugely damaging one.

Avoiding lockdown would have required a lot more effort, but the necessary knowledge, technologies and systems were all there well before the end of 2020. The problem was that none of the UK administrations were fully committed to making them work. I suggested on the BBC's *Andrew Marr Show* in March 2021 that we should regard going into lockdown as a failure of policy in the same way that we regarded the NHS being overwhelmed as a failure of policy. If governments and their scientific advisors had taken that stance then we'd surely have spent far less time in lockdown.

The third error was to manage this public health emergency as if it were *only* a public health emergency, as if it were only about the burden of death and disease caused by novel coronavirus. It wasn't. The pandemic had huge ramifications for health care provision beyond Covid-19, for mental health, for education, for the economy and for the well-being of society. We needed to manage all those harms at the same time, yet the dice were always loaded in favour of suppressing novel coronavirus at – almost literally – any cost.

Part of the fault lay with the make-up of the scientific advisory committees. The advisory system was dominated by clinicians and public health specialists who weren't looking at the bigger picture – they weren't asked to do so and weren't competent to do so if they had been. Which is why they kept recommending lockdown.

We didn't hear many other voices. The loudest voices in opposition to lockdown in the UK came from educationalists concerned about the long-term impact of closing schools on our children. We didn't hear enough from economists, especially those who could speak to the long-term impact of the damage caused to the economy, to health care provision and to education.

Nor did we hear enough from ethicists. Why wasn't there a much noisier debate about the ethics of the UK's pandemic response? As a parent, I am deeply uncomfortable with the fact that our strategy to tackle novel coronavirus did such serious harm to children and young adults. We deprived them of their education, jobs and normal existence, as well as damaging their future prospects and leaving them to inherit a record-breaking mountain of public debt. All this to protect the NHS from a disease that is a far, far greater threat to the elderly, frail and infirm than to the young and healthy, an objective that could have been achieved by other means. If there was a justification for burdening the younger generations in this way then I'd have liked to have heard it. To me, it seems morally wrong.

Once novel coronavirus took hold it was inevitable that 2020 would be an exceptionally difficult year, but there were options at every turn. If we'd trusted ourselves, our data, our systems and our science then I've no doubt we'd have made better choices. Lives and livelihoods could have been saved, lockdowns largely avoided and far less damage done. Instead, we were mesmerised by the once-in-a-century scale of the emergency and succeeded only in making a global crisis even worse. In short, we panicked.

At the beginning of this book I listed the things I did not expect to happen during the pandemic. I did not expect that elementary principles of epidemiology would be misunderstood and ignored, that tried-and-trusted approaches to public health would be pushed aside, that so many scientists would abandon their objectivity, or that plain common sense will be a casualty of the crisis. Yet – as I've explained – these things did happen, and we have all seen the result.

I didn't expect the world to go mad. But it did.

CHAPTER 18

DISEASE X

I spent two cold, grey days in January 2017 at a World Health Organization meeting in Geneva. I was one of a number of experts brought in from all over the world to advise on viral pandemic threats that should be high priorities for scientific research. After lengthy discussions, our committee came up with a sensible-sounding list that included Ebola, Lassa fever, MERS and SARS.

Myself and a few others present – including Peter Daszak, Director of Ecohealth Alliance, a New York-based non-profit organisation – wanted to add one more item, and we lobbied hard for it to be included on the list. We thought the next pandemic was just as likely to be caused by a virus that we didn't even know about yet. We succeeded in persuading the committee and the concept of 'Disease X' was born.

The next pandemic threat

The committee's conclusions were published on the World Health Organization website and received plenty of attention. The following year the committee took the Disease X concept one step further by thinking more about the kind of disease it was most likely to be. They came up with a small number of suggestions that included *highly pathogenic coronaviral diseases other than MERS and SARS.* It is hard to imagine a more accurate prediction: no-one can say the world wasn't warned. The World Health Organization's recommendations for further research didn't have much impact though. Most of the world remained fixated on flu.

It's right that we should be concerned about flu. There was an influenza pandemic – the swine flu pandemic – as recently as 2009. This involved a relatively mild strain of a type called H1N1 and the public health impact was, by pandemic standards, quite minor. One of the things that kept swine flu in check was that many older people had been exposed to related strains that gave them at least partial immunity. That was fortunate: Spanish flu – which killed around fifty million people worldwide in 1918–20 – was also caused by a H1N1 strain. By grim coincidence, a new H1N1 strain – of unknown pathogenicity – was reported from China in 2020, but has not spread widely so far.

There are even more lethal strains of influenza, collectively known as bird flus. Fortunately, these strains do not spread between humans, though scientists have been concerned for a long time that they might evolve the ability to do so. If that happened we could be facing something at least as bad as Spanish flu. This is why so much effort had gone into preparing for an influenza pandemic and we need to stay prepared.

Coronaviruses might have played second fiddle to influenza in the past, but that's all changed now. We learned very quickly that novel coronavirus is a close relative of the SARS coronavirus, so close it is considered to be a strain of the same virus species. SARS was considerably more lethal than novel coronavirus – it had a case fatality rate of 11%, more than ten times higher. It's fair to say that the world had a narrow escape in 2003 when an incipient SARS pandemic was halted and the virus was eradicated – one of the World Health Organization's proudest moments.

We were not so fortunate in 2020. This means that there has been a near-pandemic and a full-blown pandemic caused by SARS-like viruses in less than two decades. You do not need to be an expert on emerging infectious diseases to work out that we should already be worrying about SARS-3.

We must not, however, fall into the trap of preparing only for another pandemic caused by a SAR-like coronavirus or a new strain of influenza. Public health has a history of preparing for the pandemic we've just had, not the one we have next. The report of that World Health Organization meeting in Geneva in 2017 was intended to make the point that the next pandemic virus could be something quite different and present distinct challenges.

My research team has been working on the 'what next?' problem for over twenty years and we've made some progress. Back in 2001 we published the first systematic study showing that the majority of recently emerged infectious diseases have been viruses acquired from animals.

The technical term for an infection acquired from non-human animals is a zoonosis. SARS, Ebola and novel corona-virus are all viruses of zoonotic origin. Two or three new ones are found in humans in an average year, but most of them are only a minor threat to public health. Scientists are still arguing about how many potential new human viruses are still 'out there', but we do agree that it helps to try to narrow down which ones we should be most concerned about and we are beginning to find ways of doing that.

In 2016 my team published a list of around thirty zoonotic viruses that were rare but we knew could spread in human populations. We deemed these to be significant epidemic risks. While we were doing this work three viruses on the list – Chikungunya and Zika (both mosquito-borne viruses) and Ebola – all erupted into major epidemics, so our approach seems to work. All the viruses on the World Health Organization's research priorities list were already on ours, so we were in good agreement with our colleagues too.

As well as identifying known viruses that are a pandemic threat, we are also thinking hard about Disease X. In 2020 – building on earlier work by Peter Daszak and others – we published a map of the world showing where new human viruses had been discovered and where we estimated the next ones were most likely to appear. All of the eighteen new human viruses reported in the preceding decade were discovered in the locations we designated 'very high risk' or 'high risk'. Wuhan, China, was one of those locations.

In 2019 we published a pre-print reporting an analysis of the phylogenies of almost two thousand different viruses based on their genome sequences. We reported that new human viruses with epidemic potential – candidates for Disease X – are most likely to come from lineages that already contained viruses with epidemic potential, but emerge as separate introductions into humans from an animal source. Again, our approach works: novel coronavirus fits that pattern perfectly, it is closely related to another virus – SARS – that has epidemic potential but it jumped into humans independently, probably from horseshoe bats.

Emergence of new viruses

Much has been written about how new viruses emerge. I first became interested in the subject after reading Laurie Garret's book *The Coming Plague* in 1995. The emergence of any new virus – be it Ebola, MERS or anything else – is a fascinating story in its own right. No-one could predict the precise sequence of events that led to the emergence of a virus like Nipah, a story that involves bats, mangoes and a new way of farming pigs in Malaysia. I am sure that the story of the emergence of novel coronavirus will prove just as complicated, if we ever find out.

We believe we know when novel coronavirus emerged. Analysis of viral genomes supports the idea that the virus first jumped into humans in December 2019, though it is possible that it happened a month or two earlier than that. Some have proposed that it did emerge earlier and this was covered up. There are even suggestions that it had arrived in the UK by that time. I'm doubtful; this isn't a virus anyone could hide for long. December 2019 looks right to me.

That's when, but what about where? We know that many of those early cases of novel coronavirus were linked to the South China Seafood wholesale market in Wuhan, China. This is a so-called 'wet' market, selling a wide variety of live animals. The market is located just ten kilometres from the Wuhan Institute of Virology, China's leading coronavirus research laboratory. Coincidences happen and there may be no link between the virus outbreak and the lab, but this is a big one and I doubt China will ever throw off the suspicion that there was a connection.

Early in 2021 the World Health Organization dispatched a team of scientists to Wuhan to investigate. The report they published in February of that year received – to put it politely – a mixed reception. I accept that no-one could describe the report as definitive, but it does contain a lot of helpful information and some interesting analysis.

The team's preferred explanation for the emergence of novel coronavirus was that it was a spill-over event involving contact with an infected animal outside the laboratory, perhaps in the wet market, which is quite possible. Another suggestion was that it might have been brought into China in frozen food, which seems highly improbable. The report confirmed that there was no evidence that the virus was a product of genetic engineering – we'd see the signature of that in the genome, and we don't (though the idea still re-surfaces from time to time). In its conclusion, the report went further by more or less dismissing the notion that the Institute of Virology was directly involved in any way at all.

That doesn't necessarily get the Institute off the hook though. An intriguing hypothesis was put forward by *The Sunday Times* in July 2020. They reported that researchers from the Wuhan Virology Institute sometimes visited bat colonies in southern China to collect samples. They suggested that, despite whatever precautions were taken, one of the researchers had been exposed to novel coronavirus during one of these visits and brought it back to Wuhan. This seems a plausible scenario but, as I said, we may never know for certain. We know all too well what happened afterwards.

Thinking about next time

What happened afterwards was a disaster, but was far from the worst pandemic that the world could have faced. I explained in Chapter 1 that the scale, speed and severity of a pandemic are largely determined by R0 (the basic reproduction number), the generation time and the infection fatality rate.

Novel coronavirus isn't exceptional by any of those measures. There are viruses with much bigger values of R0: measles, for example. There are viruses with faster generation times, such as influenza. There are viruses with far worse infection fatality rates; Ebola is one. It's easy to imagine a scenario where even the harshest measures we took in 2020–21 will not be enough. What would we do then?

That's a question we need to start thinking about now. As our collective experience of 2020 has illustrated so starkly, every one of us has a profound interest in making sure that we have a good answer. Our governments and international agencies need to have the knowledge, the tools and the systems in place to respond quickly and effectively in the face of future threats from emerging infectious diseases. However, I won't conclude with a grand vision for meeting that challenge – that would be a whole new book in its own right. I shall end with a personal story.

In April 2020 I was asked by Emma Barnett on Radio 5 Live how it felt to see the crisis that I had been concerned about for so many years playing out in real life on a global scale. In effect, Emma was asking how did it feel to have been right all along. I made some vague reply, but I was consciously dodging the question. Here's how it truly felt.

In Chapter 1 I explained how I came to write a series of e-mails to the Chief Medical Officer of Scotland in January 2020. One of these referred to a hot-off-the-press estimate of the infection fatality rate for the new virus, just below 5%. This virus was highly transmissible – much more so than flu – and if one in twenty of those infected were going to die then we were facing a global catastrophe. I suspected that 5% was an overestimate. On the other hand, I knew that the new virus was closely related to SARS, which was even more deadly, so I couldn't be sure. On the evening of January 24th I was talking this through with my wife, broke off in mid-sentence, put my arms around her, and burst into tears. That's how it truly felt.

The first year of the novel coronavirus pandemic was a desperate time for humanity, but thankfully it wasn't as bad as I'd feared that January night in Edinburgh. This won't be the end of the story though. There will be another pandemic sooner or later and that one may well be worse. We need to be prepared.

ACKNOWLEDGEMENTS

Science is all about teamwork and I had the good fortune to be leading an outstanding team during the pandemic. Epigroup – as we're known within the University of Edinburgh – got through an extraordinary amount of work in less-than-ideal circumstances and often under tremendous pressure. I am extremely grateful to all of them: Feifei Zhang, Alex Morgan, Giles Calder-Gerver, Jordan Ashworth, Bram van Bunnik, Lu Lu, Seth Amanfo, Roo Cave, Shengyuan Zhao, Miles McGibbon, Tara Wagner-Gamble, Alistair Morrison, Miranda Ferguson, Lexy Huber, Samuel Haynes, Meghan Perry, Hannah Lepper, Miriam Karinja, Felix Stein, Stefan Rooke, Abby Peters, Donald Smith, Thomas Dalhuisen, Bryan Wee and Kath Tracy. I am equally grateful for the help we received from Stella Mazeri, Paul Bessell, Margo Chase-Topping and Camille Simonet.

I'd also like to thank my University of Edinburgh colleagues Aziz Sheikh, Francisca Mutapi, Helen Brown, Geoff Banda, Samantha Lycett and Andrew Rambaut and, from further afield, Chris Robertson, Graham Medley, Paul Kellam, Mike Parker and Chris Dye.

The University of Edinburgh and the Usher Institute supported my team, and did what they could to spare me from the day-to-day vexations of academic life while I was concentrating on pandemic response work.

I learned a huge amount from my colleagues on the science advisory committees. The expertise of SPI-M was immensely impressive and the committee members and their teams put in an exceptional effort for well over a year. Members of the Scottish Covid-19 Advisory Group did the same. Throughout the pandemic, the dedication and commitment of everyone working in the scientific advisory bodies – and no less the government officials supporting them – was plain to see.

Fiona Fox, Fiona Lethbridge and their colleagues at the Science Media Centre worked just as tirelessly to make sure that both scientific advances and scientific uncertainties were properly explained to the press. I greatly valued their consistent support and encouragement of my own efforts at science communication, culminating in this book.

David McKie and Callum Anderson took care of legal matters promptly and effectively whenever they arose. Thank you for taking a big weight off my shoulders.

Without the advice, support and encouragement of Dorothy Crawford I doubt this book would ever have seen the light of day. I was also heartened by Matt Ridley's kind enthusiasm for the project and delighted that he agreed to write the Foreword.

I am grateful to Robbie Guillory – my agent – for seeing potential in a raw first draft of this book and to Bob Davidson – my editor – for taking a chance on a novice author.

Finally, my 2020 was made far more tolerable than it might have been by a steady stream of supportive words not only from family, friends and colleagues but also from long-lost friends, distant acquaintances, scientists I'd never met and complete strangers. Thank you all.

ABBREVIATIONS

ACE-2:   angiotensin-converting enzyme 2 – the cell surface molecule that novel coronavirus and SARS use as a receptor to gain entry to a human cell

AIDS: acquired immune deficiency syndrome, caused by HIV

BSE: bovine spongiform encephalopathy, otherwise known as mad cow disease

CMO: Chief Medical Officer

COG-UK: Covid-19 Genomics UK consortium

CORSAIR: Covid-19 Rapid Survey of Adherence to Interventions and Responses – a research project on behaviour during the UK novel coronavirus epidemic

CSA: Chief Scientific Advisor

DELVE: Data evaluation and learning for viral epidemics – an initiative of the Royal Society of London contributing to the evidence base available to policy-makers

DHSC: Department of Health and Social Care

ELISA:	enzyme-linked immunosorbent assay – the method underlying some of the diagnostic tests
GDP:	gross domestic product
H1N1:	strain of the influenza A virus
HIV:	human immunodeficiency virus
ME:	myalgic encephalomyelitis – also known as chronic fatigue syndrome
MERS:	Middle East Respiratory Syndrome – a disease caused by a relative of novel coronavirus
MHRA:	Medical and Healthcare products Regulatory Agency
MIS-C:	multisystem inflammatory syndrome in children – a serious complication of infection with novel coronavirus in children
MMR:	measles, mumps and rubella – the three viruses that the MMR vaccine protects against
MP:	Member of Parliament, referring to the UK Parliament in Westminster
mRNA:	messenger ribonucleic acid – instructions that tell a cell how to make a protein
MRSA:	methicillin-resistant *Staphylococcus aureus* – a common hospital-acquired bacterial infection
MSP:	Member of the Scottish Parliament
NERVTAG:	New and Emerging Respiratory Virus Threats Advisory Group – a UK government advisory committee

NHS: National Health Service

NICE: National Institute for Health and Care Excellence

PM: Prime Minister of the UK

PPE: personal protective equipment

QALY: quality-adjusted life years – a measure of public health burden or benefit

REACT: real-time assessment of community transmission – a novel coronavirus surveillance project based at Imperial College London

RT-PCR: reverse transcription polymerase chain reaction – a laboratory technique used in diagnostic testing

SAGE: Scientific Advisory Group for Emergencies – the top UK government advisory committee during the pandemic

SARS: severe acute respiratory syndrome – caused by the SARS coronavirus, a close relative of novel coronavirus

SPI-B: Scientific Pandemic Insights Group on Behaviours – a DHSC advisory committee reporting to SAGE

SPI-M: Scientific Pandemic Influenza Group on Modelling – another DHSC advisory committee reporting to SAGE

TTI: test, trace and isolate

VOC: variant of concern, referring to variants of novel coronavirus

BIBLIOGRAPHY

Studies on novel coronavirus are being published every day. This book is not a comprehensive review of the science of novel coronavirus and the bibliography includes only selected publications that were available at the relevant time. As more research is done, our understanding of the virus will grow and perhaps change.

The basic statistics that I refer to throughout this book – such as number of novel coronavirus cases in the UK or deaths across the world – come from a number of easily accessed sources: DHSC website; Scottish Government website; World Health Organization website; Office for National Statistics website; the REACT survey; Google mobility data.

For timelines of national and international responses I used: Wikipedia's convenient compilation (though I advise double-checking cited sources); SAGE meeting minutes; Scottish Covid-19 Advisory Group papers; World Health Organization website.

Chapter 1

The World Health Organization's Covid-19 update on January 23rd 2020 was published in the News section of their web site.

Taiwan's response is described in the paper 'Potential lessons from the Taiwan and New Zealand health responses to the COVID-19 pandemic' by Jennifer Summers and colleagues published in *The Lancet Regional Health – Western Pacific* in October 2020.

The UK's 'Pandemic flu response plan' was published by Public Health England on the gov.uk website in 2014.

Chapter 2

Louise Taylor, Sophia Latham and I published our paper 'Risk factors for the emergence of human viruses' in *Philosophical Transactions of the Royal Society B* in 2001.

The International Committee on the Taxonomy of Viruses published their name for the new virus in the journal *Nature Microbiology* in March 2020.

Peng Zhou and colleagues published a pre-print identifying ACE-2 as the cell receptor used by novel coronavirus on the bioRχiv server in January 2020.

Laetitia Canini and I, together with colleagues at the Pirbright Institute, published the paper 'Timelines of infection and transmission dynamics of H1N1pdm09 in swine' in the journal *PLOS Pathogens* in 2020.

Muge Cevik and colleagues published a review of 'SARS-CoV-2, SARS CoV, and MERS Co-V viral load dynamics, duration of viral shedding, and infectiousness' in *Lancet Microbe* in November 2020.

Both the NHS and the CDC list the symptoms of Covid-19 on their web sites.

Chapter 3

Oxford University Press published Roy Anderson and Bob May's hugely influential book *Infectious Diseases of Humans Dynamics and Control* in 1991.

Andrew Wakefield and colleagues published a paper alleging side-effects of the MMR vaccine in *The Lancet* in 1998. It was retracted in 2010 but is still accessible on-line.

Imperial College's 'Report 9 – Impact of non-pharmaceutical interventions (NPIs) to reduce COVID-19 mortality and healthcare demand' was published on March 16th 2020 and is available on the MRC Centre for Global Infectious Disease Analysis website.

Sunetra Gupta and colleagues published their pre-print 'The impact of host resistance on cumulative mortality and the threshold of herd immunity for SARS-CoV-2' on the medRχiv server in July 2020.

Chapter 4

A technical report on the 'Reproduction number (R) and growth rate (r) of the COVID-19 epidemic in the UK' was published on the Royal Society of London's website in August 2020.

A paper on 'Mortality due to a second wave of COVID-19 in Scotland' is available on the advisory group evidence papers (October 2020) page of the Scottish Government web site.

Seth Flaxman and colleagues published their paper on 'Estimating the effects of non-pharmaceutical interventions on COVID-19 in Europe' the journal *Nature* in June 2020.

Simon Wood's paper 'Inferring UK COVID-19 fatal infection trajectories from daily mortality data: were infections already in decline before the UK lockdowns?' was published in the journal *Biometrics* in March 2021.

A selection of models of epidemic responses – with contributions from my own team – were included in a special edition of *Philosophical Transactions of the Royal Society B* 'Modelling that shaped the early COVID-19 pandemic response in the UK' published in May 2021.

Marc Lipsitch published his article 'Navigating the Covid-19 pandemic' on the STAT website in April 2020.

Chapter 5

Public Health England publish their reports of 'Surveillance of influenza and other respiratory viruses in the UK' on the gov.uk website.

Alvina Lai and colleagues published their paper 'Estimated impact of the COVID-19 pandemic on cancer services' in the medical journal *BMJ Open* in November 2020.

Office for National Statistics data on home deaths in 2020 can be found on their web site.

Rachel Mulholland and colleagues published their paper 'Impact of COVID-19 on accident and emergency attendances and planned hospital admission in Scotland' in the *Journal of the Royal Society of Medicine* in October 2020.

A Nuffield Trust report 'Hospital bed occupancy' (updated in June 2021) is available on their website.

Rory O'Connor and colleagues published a report on 'Mental health and well-being during the COVID-19 pandemic' in the *British Journal of Psychiatry* in October 2020.

An Office for National Statistics survey of depressive symptoms is included in a report 'What we have learned in the past month (May 2021)', available on their website.

A report 'Paediatric mortality related to pandemic influenza A H1N1 infection in England' was published in *The Lancet* in 2010.

Russell Viner and colleagues published a review of 'School closure and management practices during coronavirus outbreaks including COVID-19' in *Lancet Child & Adolescent Health* in April 2020.

The Royal Society of London's DELVE initiative published a report 'Balancing the risks of pupils returning to schools' on the github website in July 2020.

Data on excess mortality and morbidity, the burden of long covid and GDP are available on the Office for National Statistics website.

Chapter 6

Derek Chu and colleagues published their review of 'Physical distancing, face masks, and eye protection to prevent person-to-person transmission of SARS-CoV-2 and COVID-19' in *The Lancet* in June 2020.

The World Health Organization published a report 'Transmission of SARS-CoV-2: implications for infection prevention precautions' on their website in July 2020.

Public Health England published a report 'Factors contributing to the risk of SARS-CoV2 transmission in various settings' on the gov.uk website in December 2020.

Tommaso Bulfone and colleagues published a review of 'Outdoor transmission of SARS-CoV-2 and other respiratory viruses' in the *Journal of Infectious Diseases* in November 2020.

The London School of Hygiene and Tropical Medicine's novel coronavirus outbreak database is available on the covid-19settings.blogspot.com website.

In 1997, I and twelve colleagues published a paper 'Heterogeneities in the transmission of infectious agents' in *Proceedings of the National Academy of Sciences USA*.

You can find an account of Mary Mallon's life story published in 2013 by Filio Marineli and colleagues in the *Annals of Gastroenterology*.

Kaiyuan Sun and colleagues published their paper 'Transmission heterogeneities, kinetics, and controllability of SARS-CoV-2' in the journal *Science* in January 2021.

Chris Wymant and colleagues published an analysis of 'The epidemiological impact of the NHS COVID-19 app' in the journal *Nature* in May 2021.

Christophe Fraser and colleagues explain the importance of sigma in their paper 'Factors that make an infectious disease outbreak controllable' published in *Proceedings of the National Academy of Sciences USA* in 2004.

Test, Trace and Isolate performance statistics are available on the gov.uk website.

The first CORSAIR report 'Adherence to the test, trace and isolate system' was published on the medRχiv server in October 2020.

Daisy Fancourt and colleagues published a letter on "the Cummings effect" in *The Lancet* in August 2020.

Chapter 7

Fei Zhou and colleagues published their study 'Clinical course and risk factors for mortality of adult inpatients with COVID-19 in Wuhan, China' in *The Lancet* in March 2020.

David Spiegelhalter's paper 'Use of "normal" risk to improve understanding of dangers of covid-19' was published in the *British Medical Journal* in September 2020.

The QCOVID algorithm was published in the same journal in October 2020.

A study of risk factors called OpenSAFELY was published in *Nature* in July 2020.

Office for National Statistics data on age of death due to novel coronavirus and on deaths in people with and without pre-existing conditions are available on their website.

Erola Pairo-Castineira and colleagues published their paper 'Genetic mechanisms of critical illness in COVID-19' in *Nature* in December 2020.

A SAGE paper 'Contribution of nosocomial infections to the first wave' by the London School of Hygiene and Tropical Medicine and Public Health England was published in January 2021 on the gov.uk website.

The *British Medical Journal* published a news item 'COVID-19: an extra 1.7 million people in England are asked to shield' in February 2021.

My team's paper 'Segmentation and shielding of the most vulnerable members of the population as elements of an exit strategy from COVID-19 lockdown' was originally published as a pre-print on the medRχiv server in May 2020 and subsequently in *Philosophical Transactions of the Royal Society B*.

The concept of 'ubuntu' was explained in a commentary in *BMJ Global Health* in July 2020.

Chapter 8

Joel Mossong and colleagues published their paper 'Social contacts and mixing patterns relevant to the spread of infectious diseases' in the journal *PLOS Medicine* in 2008.

A review of 'Severe acute respiratory syndrome (SARS) in neonates and children' by A. M. Li and P.C. Ng was published in the journal *Archives of Disease in Childhood* in 2005.

Haiyan Qiu and colleagues published their study 'Clinical and epidemiological features of 36 children with coronavirus disease 2019 (COVID-19) in Zhejiang, China' in *Lancet Infectious Diseases* in March 2020.

A report 'COVID-19 in schools – the experience of NSW' was first published in April 2020 on the NCIRS Australia website.

Data on the risk to schoolteachers is included in the SAGE paper 'Update on children, schools and transmission' published in January 2021 on the gov.uk website.

Public Health England published a paper 'SARS-CoV-2 infection and transmission in educational settings' in *Lancet Infectious Diseases* in March 2021.

Public Health England's report on the Leicester outbreak in June 2020 is available on the gov.uk website.

An Office for National Statistics survey of novel coronavirus infection by profession dated November 2020 is available on their website.

Chapter 9

Olivier Vandenberg and colleagues published their paper 'Considerations for diagnostic COVID-19 tests' in *Nature Reviews Microbiology* in October 2020.

A World Health Organization report 'Considerations for implementing and adjusting public health and social measures in the context of COVID-19' is available on their website.

The ZOE symptoms survey can be accessed on the ZOE COVID study website.

Performance data for NHS Test and Trace are published on the gov.uk website.

A Public Accounts Committee report on 'COVID-19: Test, track and trace' was published in March 2021 on the UK Parliament website.

The European CDC publishes novel coronavirus testing data across Europe on their website.

Chapter 10

The Royal Society of London's DELVE initiative published a report 'Economic aspects of the COVID-19 crisis in the UK' on the github website in August 2020.

Walter Dowdle set out 'The principles of disease eradication and elimination' in the *MMWR* journal in 1999.

The National Audit Office's report 'The 2001 Outbreak of Foot and Mouth Disease' is available on their website.

Public Health England's report 'Investigation into the effectiveness of "double testing" travellers incoming to the UK for signs of COVID-19 infection' was published as a SAGE paper on the gov.uk website in September 2020 and the Animal and Plant Health Agency's report 'The risk of introducing SARS-CoV-2 to the UK via international travel in August 2020' was published on the medRχiv server in the same month.

Timothy Russell and colleagues published their paper 'Effect of internationally imported cases on internal spread of COVID-19' in *Lancet Public Health* in December 2020.

Chapter 11

The Academy of Medical Sciences report 'Preparing for a Challenging Winter 2020/21' was published on their website in July 2020.

Ellen Brooks-Pollock and colleagues published a pre-print on 'High COVID-19 transmission potential associated with re-opening universities' on the medRχiv server in September 2020.

Chapter 12

COG-UK's paper 'Establishment and lineage dynamics of the SARS-CoV-2 epidemic in the UK' was published in the journal *Science* in February 2021.

COG-UK's paper 'Epidemic waves of COVID-19 in Scotland' was published on their website in December 2020.

Emma Hodcroft and colleagues published a pre-print 'Emergence and spread of a SARS-CoV-2 variant through Europe in the summer of 2020' on the medRχiv server in March 2021.

Rob Challen and colleagues published their paper 'Risk of mortality in patients infected with SARS-CoV-2 variant of concern 202012/1' in the *British Medical Journal* in March 2021.

Dan Frampton and colleagues published their paper 'Genomic characteristics and clinical effect of the emergent SARS-CoV-2 B.1.1.7 lineage in London, UK' in *Lancet Infectious Diseases* April 2021.

Vaughn Cooper wrote an article 'The coronavirus variants don't seem to be highly variable so far' in *Scientific American* magazine in March 2021.

Chapter 13

Clinical trial data for the Moderna and Pfizer vaccines were published in separate papers in *New England Journal of Medicine* in December 2020 and for the AstraZeneca vaccine in *The Lancet* in the same month.

The EAVE project's paper 'Interim findings from first-dose mass COVID-19 vaccination roll-out and COVID-19 hospital admissions in Scotland' was published in *The Lancet* in May 2021.

Reports from Public Health England's SIREN study of vaccine effectiveness are available on the gov.uk website.

Jackson Turner and colleagues published their paper 'SARS-CoV-2 infection induces long-lived bone marrow plasma cells in humans' in the journal *Nature* in May 2021.

Shabir Madhi and colleagues published their paper 'Efficacy of the ChAdOx1 nCoV-19 Covid-19 vaccine against the B.1.351 variant' in *New England Journal of Medicine* in March 2021.

Medicines and Healthcare products Regulatory Agency advice on the safety of the AstraZeneca vaccine is available on the gov.uk website.

Chapter 14

Annakan Navaratnam and colleagues published their paper 'Patient factors and temporal trends associated with COVID-19 in-hospital mortality in England' in *Lancet Respiratory Medicine* in February 2021.

Nature published a news feature 'The coronavirus is here to stay – here's what means' in February 2021.

An example of the SPI-M modelling studies used to guide policy in the first half of 2021 is the report 'Road map scenarios and sensitivity: steps 3 and 4' by the University of Warwick team, available on the gov.uk website.

Rachel Wilf-Miron and colleagues published an article on 'The "green pass" proposal in Israel' in the journal *JAMA* in March 2021.

The JUNIPER consortium published a pre-print 'Early epidemiological signatures of novel SARS-CoV-2 variants' on the medRχiv server in June 2021.

Chapter 15

The Institute for Health Metrics and Evaluation's report 'Estimation of excess mortality due to COVID-19' was published on their website in May 2021 and the World Health Organization's report 'World Health Statistics 2021' is available on their website.

Michael Baker and colleagues published a report 'Successful elimination of Covid-19 transmission in New Zealand' in *New England Journal of Medicine* in August 2020.

Anders Tegnell published an article 'The Swedish public health response to COVID-19' in the *Journal of Pathology, Microbiology and Immunology* in February 2021.

The World Health Organization Africa Region's novel coronavirus Situation Reports are available on their web site.

A World Bank report 'Assessing the Economic Impact of Covid-19 and Policy Responses in Sub-Saharan Africa' was published on their website in April 2020.

Save the Children published a report 'COVID-19 impacts on African children' in June 2020 and UNICEF published a report 'COVID-19: a catastrophe for children in sub-Saharan Africa' in November 2020, available on their websites.

The World Health Organization's recommendations for international traffic in relation to COVID-19 published in February 2020 are available on their website.

The Independent Panel for Pandemic Preparedness & Response report 'COVID-19: Make it the Last Pandemic' was published in May 2021 and is available on their website.

Chapter 16

A report by the Department for Business, Energy & Industrial Strategy on the 'UK Research Base 2016' is available on the gov.uk website.

Richard Doll and Bradford Hill's seminal paper 'Smoking and carcinoma of the lung' was published in the *British Medical Journal* in 1950.

Naomi Oreskes discusses the link between science and practice in her book *Why Trust Science?* published in 2019 by Princeton Press.

Ben Goldacre's book *Bad Science* was published in 2008 by Fourth Estate.

A report 'Research and development expenditure by the UK government: 2018' is available on the Office for National Statistics website.

The Foresight report 'Infectious diseases: preparing for the future' was published in 2006 and is available on the gov.uk website.

The Life of Voltaire by Evelyn Beatrice Hall (pen name S. G. Tallentyre) was published in 1903 by Smith, Elder & Co.

Chapter 17

Stephen Kissler and colleagues published their paper 'Projecting the transmission dynamics of SARS-CoV-2 through the postpandemic period' in the journal *Science* in May 2020.

The Great Barrington Declaration statement on "focused protection" is available on the website gbdeclaration.org.

Jeanne Lenzer and Shannon Brownlee published an article 'The COVID science wars' in the magazine *Scientific American* in November 2020.

Julia Mikolai and colleagues published their paper 'Intersecting household-level health and socio-economic vulnerabilities and the COVID-19 crisis' in the journal *SSM – Population Health* in December 2020.

Sonja Rasmussen and colleagues wrote about 'Precision public health as a key tool in the COVID-19 response' in the journal *JAMA* in August 2020.

Daniel Kahneman's book *Thinking, Fast and Slow* was published in 2011 by Farrar, Straus and Giroux.

Chapter 18

The World Health Organization's '2018 annual review of diseases' is available on their website.

Laurie Garrett's book *The Coming Plague* was published in 1994 by Farrar, Straus and Giroux.

Dorothy Crawford describes the emergence of Nipah virus in her book *Viruses: a very short introduction* published by Oxford University Press in 2011.

I and colleagues published our paper 'Assessing the epidemic potential of RNA and DNA viruses' in *Emerging Infectious Diseases* in 2016.

Feifei Zhang, I and colleagues published our paper 'Global discovery of human-infective RNA viruses' in *PLOS Pathogens* in November 2020.

Lu Lu, I and colleagues published a pre-print 'Evolutionary origins of epidemic potential among human RNA viruses' on the bioRχiv server in 2019.

The Times Insight Investigation team published a discussion of potential origins of novel coronavirus in *The Sunday Times* in July 2020 (available on their website but behind a pay wall).

The World Health Organization's report 'Global study origins of SARS-CoV-2: China part' was published in April 2021 and is available on their website.

There are several books on the Spanish influenza pandemic of 1918, including Laura Spinney's *Pale Rider* published in 2017 by Cape.

INDEX